Lecture Notes in Civil Engineering

Volume 60

Lecture Notes in Civil Engineering (LNCE) publishes the latest developments in Civil Engineering—quickly, informally and in top quality. Though original research reported in proceedings and post-proceedings represents the core of LNCE, edited volumes of exceptionally high quality and interest may also be considered for publication. Volumes published in LNCE embrace all aspects and subfields of, as well as new challenges in, Civil Engineering. Topics in the series include:

- Construction and Structural Mechanics
- Building Materials
- Concrete, Steel and Timber Structures
- Geotechnical Engineering
- Earthquake Engineering
- Coastal Engineering
- Ocean and Offshore Engineering; Ships and Floating Structures
- Hydraulics, Hydrology and Water Resources Engineering
- Environmental Engineering and Sustainability
- Structural Health and Monitoring
- Surveying and Geographical Information Systems
- Indoor Environments
- Transportation and Traffic
- Risk Analysis
- Safety and Security

To submit a proposal or request further information, please contact the appropriate Springer Editor:

- Mr. Pierpaolo Riva at pierpaolo.riva@springer.com (Europe and Americas);
- Ms. Swati Meherishi at swati.meherishi@springer.com (Asia—except China—and Australia/NZ);
- Ms. Li Shen at li.shen@springer.com (China).

Indexed by Scopus

More information about this series at http://www.springer.com/series/15087

Arun Sharma · Radha Goyal · Richie Mittal
Editors

Indoor Environmental Quality

Select Proceedings of the 1st ACIEQ

Springer

Editors
Arun Sharma
Department of Community Medicine
University College of Medical Sciences
New Delhi, Delhi, India

Radha Goyal
Indian Pollution Control Association
New Delhi, Delhi, India

Richie Mittal
Overdrive Engineering Solutions Pvt. Ltd
Noida, Uttar Pradesh, India

ISSN 2366-2557 ISSN 2366-2565 (electronic)
Lecture Notes in Civil Engineering
ISBN 978-981-15-1333-6 ISBN 978-981-15-1334-3 (eBook)
https://doi.org/10.1007/978-981-15-1334-3

This Springer imprint is published by the registered company Springer Nature Singapore Pte Ltd.
The registered company address is: 152 Beach Road, #21-01/04 Gateway East, Singapore 189721, Singapore

Foreword

It gives immense pleasure that the first Asian Conference on Indoor Environmental Quality 2019 (ACIEQ 2019) is being organized by ISHRAE, SIE and IAQA.

ACIEQ 2019 is the first of its kind conference dealing with indoor environment, which is an area less explored in India. This conference will initiate a tradition of bringing together researchers, academics, professionals and experts in Indoor Environment science, from all over the world.

This conference will particularly encourage the interaction of research students and developing academics with the more established academic community and professionals in an informal setting to present and to discuss new and current work on Indoor environment. Their contributions will help to make the Conference an outstanding one with papers contributing the most recent scientific knowledge known in the field of exposure assessment, thermal comfort, acoustics, indoor environment and Sustainable Development.

In addition to the contributed papers, three invited keynote presentations will be given by Prof. Alan Hedge, Cornell University, Prof Paolo Carrer from University of Milan and Prof. Richard D. Dear from University of Sydney representing all aspects of indoor environment ranging from building design, health and ventilation and thermal comfort will be discussed at the conference.

I wish the organizers and delegates, attending this conference, a good knowledge transfer and betterment of application based scientific temperament through this conference.

FOR A HEALTHY INDOOR ENVIRONMENT!

24 January 2019

(Hardik Shah)
PS to Hon'ble Minister

Preface

Indoor environmental quality (IEQ) plays an important role in maintaining the health and well-being of the individuals especially in urban areas as people spend almost 90% of their time indoors. IEQ is gaining importance in developing India as new infrastructures are on the rise. So, it has become imperative to strike the balance between the IEQ, ventilation, human health and productivity. Indoor air quality (IAQ) is a prime area of concern coupled with efficient lighting, adequate thermal comfort and manageable acoustics, determining the health and productivity of the urban population in India.

Keeping in mind the need to create a platform to bring all the stakeholders of IEQ together to discuss the issues, challenges and possible solutions for better IEQ, Indian Society for Heating Refrigeration and Air Conditioning Engineers (**ISHRAE**), Society for Indoor Environment (**SIE**) and Indoor Air Quality Association (**IAQA**) together had organized the first edition of Asian Conference on "Indoor Environmental Quality" which took place at Indian Aviation Academy, Vasant Kunj, New Delhi, India, on February 1 and 2, 2019. It is for the first time that an international conference on the subject of IEQ was being organized in India. The conference theme was entitled "**Habitable Built Environment—Experience the Unseen**."

For this conference, research papers were invited under four major themes, i.e., *Indoor Air Quality, Thermal Comfort, Lighting, and Acoustics*. The conference received a total of 62 abstracts of papers, out of which only 40 were selected by the scientific committee for full paper submission. On full paper call, the conference received only 22 full papers, out of which only 17 papers were accepted for oral presentation and three papers were accepted for poster presentation. Remaining oral presentations were from the invited speakers.

The proceedings of the conference (having 18 chapters in all—17 chapters are from oral presentation and one from poster presentation) are being brought to you in this book form for wider dissemination of the presentations made at the conference. It will not only be an update on IEQ research but also a trigger for research work that needs to be undertaken in Asia to tackle the challenges of IEQ head on. Included here are the papers on indoor environment quality monitoring and modeling; the influence of confounding factors like thermal comfort parameters, such as temperature and relative humidity with respect to different building types (e.g., residential, commercial, institutional); ventilation characteristics and lighting. The research papers on people's performance, productivity and behavior with respect to their exposure to various indoor air pollutants and parameters influencing the overall indoor environmental quality are also included.

New Delhi, India — Arun Sharma
New Delhi, India — Radha Goyal
Noida, India — Richie Mittal

Message

Of three types of air pollution, the ambient, household, and indoor, the last one receives little focus in India, in particular that of researchers. It is understandable given the magnitude of poor air quality that most investigators have pursued research in ambient air pollution and associated dimensions. However, with the use of new materials and compact building designs the indoor air quality has undergone sea change. Since it does not cause acute effects, and rarely ever kills, the issue has not caught the imagination of researchers and communities.

Poor indoor air quality and environment are cause of several building related illnesses that many physicians are not familiar with. One of the deadliest is Legionnaires disease that can be fatal in case of elderly and is a key health disorder resulting from poor maintenance of cooling towers. The lack of prescribed standards for Indoor Air Quality is another reason for lack of awareness about this important subject. Environmentalists and health professionals must become aware of these issues and try to understand the environment triggering an altered biological and pathological response of human body. I hope the conference will shed light on various aspects of IAQ issue and bring it to the attention of policy makers, regulators, building and construction lobbies and last but not the least HVAC engineers. I send my best wishes for the success of the conference.

(T.K. JOSHI)

T K Joshi
Fellow Collegium Mazzini

Adviser, Environmental Health
MoEFCC, GOI

Message

The health concerns due to poor indoor air quality have been increasing rapidly in India and other parts of the world. Initially, the sources of indoor pollutants were limited to fuel used for cooking and for heating purposes. This pattern has been changing due to additional sources prevalent indoors mainly from mixture of products and material used for comfortable living. These starts from apparel, perfumes, and insecticides used indoors, personal care products etc. Growth rate of personal care products including use of mosquitoes repellent are very high and consequent health complaints have also been on a serious growth path. Mere use of technologies alone will not be able to address this issue. It is important that we undertake comprehensive scientific and social awareness for people to know what is likely to cause harm. In addition, the scientific community must work with manufacturers of the products to reduce emission when used indoors. The partnership of all the stakeholders will go a long way in addressing this problem. The event "Asian Conference on Indoor Environmental Quality" will be able to bring all partners together to debate and recommend the best path forward for the India and the world. Best wishes.

28 January 2019

(Rakesh Kumar)
Director, CSIR—NEERI

Message

Air pollution has become a serious concern, particularly for health of the people. People are exposed to unacceptable levels of air pollution in the outdoor as well as indoor environment. Growing population, rapid urbanization, multiplicity and complexity of sources, inadequate capacities and pace of technology and infrastructure upgradation, lack of public awareness & participation are some of the factors that make air pollution more challenging. Many policy initiatives and interventions have been taken in the recent past—introduction of BS-IV and firm roadmap for BS-VI auto fuel and emission norms, stricter emission standards for major industrial sectors, Swatchh Bharat Mission, National Solar Mission and Ujjawala Scheme for LPG proliferation. However, most of the actions pertain to outdoor air pollution. Indoor air pollution, though reported to be a larger health issue, has got little attention.

I am glad Society for Indoor Environment, ISHRAE and IAQA, India Chapter are jointly organizing 1st Edition of Asian Conference on Indoor Environmental Quality. It would provide a platform to initiate discussion with various stakeholders (research, academia, industry, societies, government and regulatory bodies) on the subject and help setting a clear roadmap for guiding future actions, particularly with regard to research studies, policies, regulatory framework, IEC, and technology. CPCB would be happy to extend all possible support to this event and look forward to successful deliberations and outcome.

My best wishes!

(Prashant Gargava)

ACIEQ 2019 Committees

Organizing Committee

Chairman

Mr. C. Subramaniam, ISHRAE

Co-chairman

Prof. Mukesh Khare, SIE

Convener

Mr. Richie Mittal, IAQA/ISHRAE

Co-convener

Mr. Gaurav Vasudev, ISHRAE

Scientific Committee

Chairman

Dr. Prasad Modak, SIE

Co-chairman

Dr. Jyotirmaya Mathur, ISHRAE

Convener

Dr. Radha Goyal, SIE/IPCA

Co-convener

Dr. Priyanka Kulshreshtha, SIE/IPCA

Coordinator

Ms. Rakhi Ghosh, IPCA

Secretary

Mr. Barun Agarwal, IAQA/ISHRAE

Treasurer

Mr. Amit Goel, ISHRAE

Administration

Mr. Vishal Kapur, ISHRAE

Chair Speakers

Dr. Nitin Deodhar, ISHRAE
Dr. Arun Sharma, SIE
Mr. Ashish Jain, SIE/IPCA

Chair Marketing

Mr. Vikram Murthy, ISHRAE
Mr. Kedar Patki, ISHRAE

Chair Sponsorship

Mr. Ashish Rakheja, ISHRAE

Chair—Travel and Hospitality

Mr. Nitin Nayak, IAQA

Chair—Event

Mr. Anoop Ballaney, ISHRAE

Chair Social and Website

Mr. Ashu Gupta, IAQA

Chair Delegate

Mr. Kedar Patki, ISHRAE
Mr. Aakash Patel, ISHRAE

Partners

Knowledge Partners	National Environmental Engineering Institute (NEERI)
	Central Pollution Control Board (CPCB)
	Centre of Excellence for Research on Clean Air (CERCA)
	Camfil
Silver Partners	ABS Airtech Pvt. Ltd.
	Aeropure
	PureAir
	Vehicle Tracking System (VTS)
	Honeywell
	Renson
Swaroski Partners	IntelliGreen Technologies Pvt. Ltd.
	Durkee
	DRI—Innovative Air Solutions
Table Partners	HUMIDIN
	Airveda
	Smoke Solution India
	Kaiterra
	Magneto Clean Tech
	EI
Workshop Partners	TSI Incorporated
	BE—Breathe Easy
	PARTICLES PLUS, INC.
	Springer
Supporting Partners	Indian Institute of Technology, Madras
	Green Business Certification Inc. (GBCI)
	IEQ-Global Alliance
	GRIHA Council—Green Rating for Integrated Habitat Assessment
	Indian Pollution Control Association (IPCA)
	International Facility Management Association (IFMA)
	The Associated Chambers of Commerce & Industry of India (ASSOCHAM, India)
	Federation of European Heating, Ventilation and Air Conditioning Associations (REHVA)
	Environmental Management Centre (EMC)
	GREEN TELS

Contents

About the Editors

Arun Sharma is a Professor in the Department of Community Medicine and Head of the Department of Bio-statistics and Medical Informatics at the University College of Medical Sciences, India. He has done his MBBS from University of Calcutta and MD in Community Medicine from Banaras Hindu University, in addition to having done fellowships with the University of Illinois, Urbana-Champain (USA), University of Pennsylvania (USA), and the Johns Hopkins Bloomberg School of Public Health (USA). Prof Sharma's research work studies the impact of pollutants on the respiratory health of urban populations.

Radha Goyal is the Deputy Director of the Indian Pollution Control Association (IPCA), where she is responsible for all research and development related activities and projects of IPCA in area of air and waste management. Dr Goyal is in-charge of the Air Quality Management Services (AQMS) division of IPCA, specifically working on indoor air quality, in addition to being a principal investigator for a number of projects mapping air quality in Delhi NCR and assessing it's impact. She has done her PhD from the Department of Civil Engineering, Indian Institute of Technology, Delhi.

Richie Mittal is the Managing Director of OverDrive Engineering Private Limited, which specialized in energy efficient design and engineering of air conditioning and cooling systems for enclosed facilities. He has served on the Board of Directors of the Indoor Air Quality Association (USA), and is a member of many professional societies and associations.

Comparison of Indoor Air Quality for Air-Conditioned and Naturally Ventilated Office Spaces in Urban Area

Supreme Jain, Divyam Garg and Anubha Goel

Abstract Indoor air quality (IAQ) is affected by indoor pollution sources that release gases or particles into the air and depending on ventilation conditions, outdoor ambient air quality can also be contributing factor. Inadequate ventilation by limiting inflow-outflow of pollutants, proximity to field or busy road can increase indoor pollutant levels. For this study, two offices, one with natural ventilation and other with central AC system, within residential campus of IIT Kanpur were selected. Air quality in offices with varying ventilation conditions was monitored simultaneously indoor and outdoor. The aim was to compare the data for the two locations and elicit influence of ventilation conditions on indoor pollutant levels. Size segregated mass concentration of ambient aerosols and particle number concentration (PNC) were determined for indoor using cascade impactor, MOUDI and OPC, respectively. Outdoor coarse particle concentration (PM_{10}) was determined using HVS sampler. The PM concentration was found to be highest in naturally ventilated office. As per the recently released ISHRAE standards for indoor environment (ISHRAE Standard—10001:2016), minimum acceptable limits for Class C are $PM_{10} < 100\ \mu g/m^3$ and $PM_{2.5} < 25\ \mu g/m^3$. Levels of particles recorded in these office spaces were 4–6 times higher than ISHRAE acceptable limits for $PM_{2.5}$ and $PM_{10,}$ respectively. Particle number concentration (PNC) was also higher in naturally ventilated office space as concentration was more than 400,000 particles/cm^3 (<1 μm) as compared to air-conditioned office space (160,000 particles/cm^3). The I/O ratio for PM10 was found to be 0.64 and 2.93 for Office 1 and Office 2, respectively.

Keywords IAQ · Office microenvironment · Particles · Ventilation · Activity level

---

S. Jain · D. Garg · A. Goel (✉)
Department of Civil Engineering, Indian Institute of Technology, Kanpur, India
e-mail: anubha@iitk.ac.in

A. Goel
Centre for Environment Science and Engineering,
Indian Institute of Technology, Kanpur, India

A. Sharma et al. (eds.), *Indoor Environmental Quality*,
Lecture Notes in Civil Engineering 60,
https://doi.org/10.1007/978-981-15-1334-3_1

1 Introduction

Indoor air quality (IAQ) which refers to the quality of air inside and around closed structures where a living being can reside is related to the comfort and health of the occupants. IAQ is represented by the concentrations of pollutants and thermal conditions. Common issues associated with IAQ are improper or inadequately maintained heating and ventilation systems, contamination by construction materials, glues, fibreglass, particle boards, paints, chemicals, increase in number of building occupants and time spent indoors. All these issues give rise to indoor air contaminants that mainly include chemicals and organics like dust, moulds or fungi, bacteria, gases, vapours and odours. Importance of air quality indoors can be considered by the statement that people spend about 90% of their time in different indoor microenvironments.

Studies related to measurement of outdoor pollution from India are numerous [1–5] as compared to a lesser number associated with examination of indoor air quality [6–10]. Proliferation of chemical pollutants in consumer and commercial products, the tendency towards tighter building envelopes and reduced ventilation to save energy and pressures to defer maintenance and other building services to reduce costs have fostered IAQ problems in most of the buildings. Knowledge of emissions from indoor sources is increasing but more information is required to be gathered for clearer understanding of the factors which impact indoor air particles concentration.

The objective of the current study is to measure and compare the variation in particulate matter mass and number concentration in two different office buildings of different ventilation system, i.e. air-conditioned and natural, inside the residential campus of IIT Kanpur. This information will enable efficient IAQ management.

2 Methodology

2.1 Site Description

The two offices microenvironment considered in this study differed in ventilation mechanism: one was fully air-conditioned (O-1) and other naturally ventilated (O-2). Postgraduate student office O-1 is a closed air-conditioned room having a seating capacity of 24 people. Post office in the premises of IIT Kanpur (O-2) is a naturally ventilated official single-storeyed building with staff of 8–10 people and rest is customers that come for the services. It is located near to the road having greenery on all four sides with some patches of dusty floors. The Offices O-1 and O-2 were occupied throughout the working hours by 10–30 and 50–70 occupants, respectively (more details listed in Table 1).

Table 1 Details of microenvironment selected for this study

	O-1	O-2
Location	Postgraduate office inside academic building	Post Office near shopping centre
Proximity to traffic routes	Distant from busy road	Distant from busy road but close to secondary road
Land use type near site	Office space	Office space
Occupants	10–30	>15

Table 2 Instrument used in this study

Instruments	Parameters examined for	Instrument description
MOUDI (Micro-Orifice Uniform Deposit Impactors)	Size segregated mass concentration	MOUDI^{TM} Model 110: without rotator MSP Corporation, USA
OPC (Optical Particle Counter)	Particle number concentration	OPC 1.108, GRIMM Aerosol Technik GmbH & Co. KG, Germany
HVS (High Volume sampler)	PM_{10} concentration	Model No. APM 460 PUF Envirotech Instruments Private Limited, India

2.2 *Instrumentation*

Real-time air samples for five days were collected inside both offices (O-1 and 0-2) in the months of March and April 2018 in this study. The instruments used for this purpose are listed in Table 2.

The MOUDI and OPC were utilised to measure mass concentration and number concentration inside the offices whereas HVS was used outdoors to measure the ambient PM10 mass concentration. A 47 mm tissue quartz filter was used to collect size segregated particulate matter on six-stage MOUDI. The six-stage MOUDI contain 0.18, 0.32, 0.56, 1.0, 1.56 and 3.16 μm cut size nozzles. Flow discharge of 30 lpm was maintained in MOUDI using a rotameter. The filter substrates of MOUDI were cleaned with 99.99% ethanol after every run to minimise the clogging.

3 Results and Discussion

3.1 *Particle Mass Concentration Inside the Offices*

The particle mass concentration was observed to be higher inside the premises of naturally ventilated Office (O-2). Here particle concentrations were found to be

600.53 ± 132.61 μg/m^3 and 255.22 ± 79.81 μg/m^3 for PM_{10} and PM_1, respectively. These values are 30–40% higher than values recorded in the air-conditioned office O-1 where concentration was found to be 416.48 ± 45.72 μg/m^3 and 194.81 ± 32.21 μg/m^{3}for PM_{10} and PM_1, respectively. The reason for large particle mass concentration in Office 2 is its ventilation condition. The transport of outdoor particles across the building envelope (i.e. penetration) is an important physical behaviour that contributes to the particle concentration [11]. Office 2 is post office inside the Campus of IIT, Kanpur which is naturally ventilated with open windows and doors being used continuously throughout the day by visitors.

Particle size distribution in air-conditioned office (O-1). Figure 1 shows the percentage contribution of particulate matter of different size bins used for sampling on different days in Office1 (O-1). The table attached with the figure is the mass concentration in μg/m^3 of different sizes on different days. Concentration of particles in all the six stages on all five days of sampling shows consistent results. O-1 is air-conditioned office space for postgraduates next to the laboratory and remains engaged throughout the week. Frequent movement of the students and staff is very common in this microenvironment and results in regular opening and closing of gates. Also, some construction activity was taking place in the backyard of the office during sampling time raising possibility of higher loading for both coarser and finer particles. From Fig. 1, loading in finer and coarser range shows little variation. Thus, it can be inferred that even if the rooms are air-conditioned, frequent movement may lead to higher particle loading and it may not satisfy standards for indoor air quality in terms of particulate matter.

Particle size distribution in naturally ventilated post office building (O-2). Similar to Figs. 1 and 2 presents particle size distribution for post office inside IIT Kanpur. From Fig. 2, it can be seen that major contribution in particulate matter is of coarser range particles, S3 to S5 stages, which arise due to re-suspension of dust due to movement of people and bags of post that are prepared in different duration of the

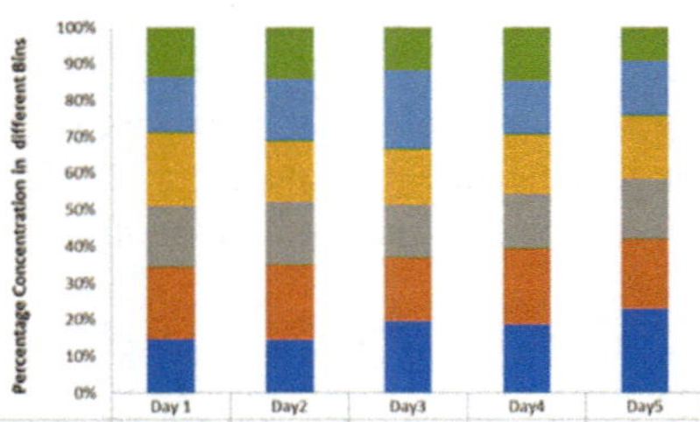

	Day 1	Day2	Day3	Day4	Day5
S8 (0.18 - 0.32 μm)	57.9	64.8	55.6	55.1	34.7
S7 (0.32 - 0.56 μm)	64.4	75.9	102.3	54.6	54.2
S6 (0.56 - 1.0 μm)	83.8	75.5	71.3	60.2	63.9
S5 (1.0 - 1.56 μm)	70.4	76.4	68.1	55.6	60.6
S4 (1.56 - 3.16 μm)	81.9	91.7	81.5	77.3	69.9
S3 (3.16 - 10 μm)	62.5	65.7	92.1	69.9	84.7

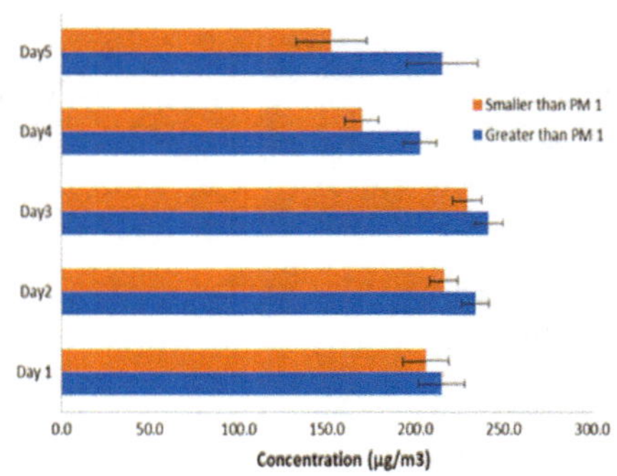

Fig. 1 Percentage contribution of different size bins to PM_{10} (Left) and distribution of particle mass larger than PM_1 and smaller than PM_1 concentration (Right) inside Office 1 (O-1)

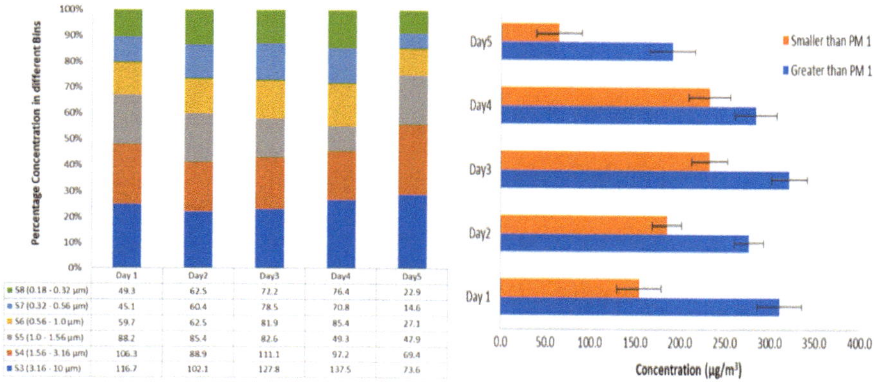

Fig. 2 Percentage contribution of different size bins to PM_{10} (Left) and distribution of particle mass larger than PM_1 and smaller than PM_1 concentration inside Office 2 (O-2)

day. Day 5 of sampling was Saturday which is holiday in the campus and shows considerably lower concentrations in comparison with the first four days. Lesser number of people visit post office on holiday affecting particle concentrations observed. The sources of finer particles in the office can be vehicular emissions and the printer used for the posts [12, 13].

3.2 *Effect of Occupant Activity on Diurnal Trends in Particle Number Concentrations Inside Office Space*

Figures 3 and 4 show the time series of particle number concentration during the sampling period in the offices. Particle sizes presented have been divided into three ranges: smaller than 1 μm, 1–3 μm and 3–10 μm. The particles seem to follow consistent trend throughout the sampling duration at both the locations.

In case of O-1, particle counts are high in the morning then decrease continuously till 2 p.m. and follow consistent trend till 5:30 pm (Fig. 3a–c). After 5:30 pm, there is sudden rise and particle count follows consistent trend for few hours. The trend followed is consistent with occupant activities. In morning, people start coming to the office, and frequent opening and closing of doors cause re-suspension of dust. This falls in the lunchtime as people start settling for rest. Again, in evening during winding up of work to and fro movement starts causing rise in the concentrations of particle.

The naturally ventilated Office (O-2) also showed the trend of particle concentration rising and falling throughout the day according to the activities of the occupants. From Fig. 4a–c, this can be seen that day starts with high particle concentration in the morning and falls continuously till 3 pm. After that particle count again increases and remains high till the completion of the sampling. Morning is the

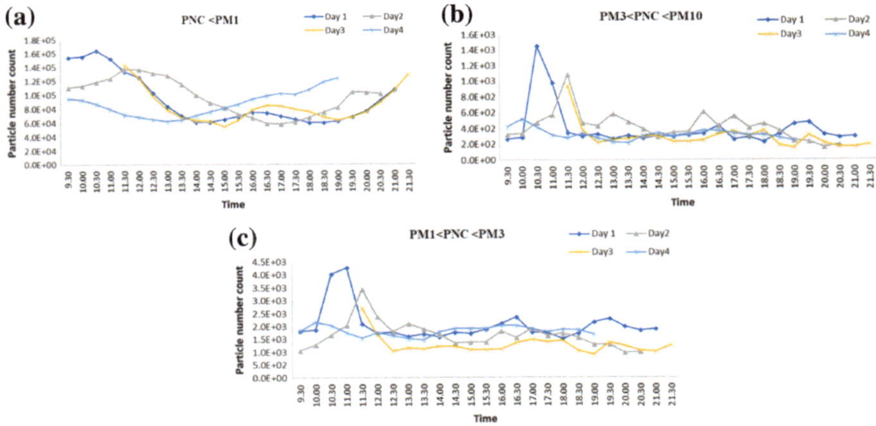

Fig. 3 Particle number concentration in particle/cm^3 inside Office 1 (O-1)

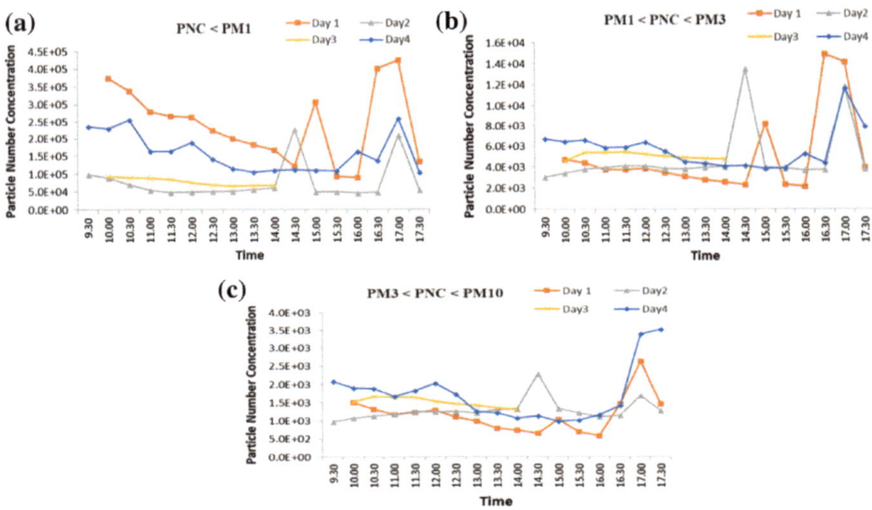

Fig. 4 Particle number concentration in particle/cm^3 inside Office 2 (O-2)

time of high activity when office workers start arriving to work. The routine bags of post to be distributed on that day are opened around 10:30 am. Till 3–3:30 pm regular posts are accepted. After 3:30 pm., bags were prepared for the posts that were collected for the day and office cleaning was conducted during the evening time.

Comparison of particle count with activity level shows that occupant movement, indoor cleaning and other activity can be the prime reason for high particle count indoor.

3.3 *Indoor-Outdoor (I/O) Relationship*

Concentration of indoor air pollutants is affected by intrusion of outdoor pollutants into the buildings and indoor sources. Indoor/outdoor (I/O) concentrations can vary largely due to many factors including locations, building design and different activities. To measure the influence of outdoor air and indoor sources on indoor air quality, I/O ratios were calculated. The I/O ratio for PM10 was found to be 0.64 and 2.93 for Office 1 (O-1) and Office 2 (O-2), respectively. For O-2, I/O ratio greater than 1 indicates that there are more particles generated indoor than outdoors. In the absence other specific source of particulate matter, human activity was seemingly the most important factor contributing to higher indoor coarse particles levels inside O-2. This has also been shown in many studies that mass movement or other human activity may lead to high particulate emission [6, 9, 14–16]. On the other hand, concentration of particulate matter inside the furnished buildings may also decrease due to sedimentation onto surface, turbulent diffusion and electrical effects. Office 1 is well furnished with respect to Office 2, supporting the reason for low I/O ratio.

4 Conclusion

The office spaces in our study have high levels of particles and do not even fall in the Class C category of the recently released ISHRAE standards for indoor environment (ISHRAE Standard—10001:2016), minimum acceptable limits for Class C category are $PM_{10} < 100\ \mu g/m^3$ and $PM_{2.5} < 25\ \mu g/m^3$. For both the office spaces, the particle concentrations exceed more than 4 times the National Ambient Air Quality Standards for ambient environment, India (NAAQS, PM_{10}: 100 $\mu g/m^3$, $PM_{2.5}$: 60 $\mu g/m^3$), which is quite alarming. It was found that the naturally ventilated open office space is quite more polluted with respect to the closed air-conditioned space.

Factors affecting the indoor air quality can be attributed to the frequency of usage of building, proximity to the roads and building type (public, private, etc.) as it controls the level of maintenance and cleanliness. Particle levels noted in indoor office spaces can be considered an environmental health threat, which reduces productivity and requires implementation of mitigation measures at the earliest. A multidisciplinary approach should be taken to mitigate the impact of pollutants inside indoor microenvironments with involvement of all stakeholders.

References

Chakraborty A, Ervens B, Gupta T, Tripathi SN (2016) Characterization of organic residues of size-resolved fog droplets and their atmospheric implications. J Geophys Res Atmos 121:4317–4332. https://doi.org/10.1002/2015JD024508

Izhar S, Goel A, Chakraborty A, Gupta T (2016) Annual trends in occurrence of submicron particles in ambient air and health risk posed by particle bound metals. Chemosphere 146:582–590. https://doi.org/10.1016/j.chemosphere.2015.12.039

Singh DK, Gupta T (2016) Source apportionment and risk assessment of PM1 bound trace metals collected during foggy and non-foggy episodes at a representative site in the Indo-Gangetic plain. Sci Total Environ 550:80–94. https://doi.org/10.1016/j.scitotenv.2016.01.037

Goyal R, Khare M (2011) Indoor air quality modeling for PM10, PM2.5, and PM1.0 in naturally ventilated classrooms of an urban Indian school building. Environ Monit Assess 176:501–516. https://doi.org/10.1007/s10661-010-1600-7

Goel A, Rathi S, Agrawal M (2018) Toxicity potential of particles caused by particle-bound polycyclic aromatic hydrocarbons (PPAHs) at two roadside locations and relationship with traffic. Environ Sci Pollut Res 25:30633–30646. https://doi.org/10.1007/s11356-018-3043-6

Jan R, Roy R, Yadav S, Satsangi PG (2017) Exposure assessment of children to particulate matter and gaseous species in school environments of Pune, India. Build Environ 111:207–217. https://doi.org/10.1016/j.buildenv.2016.11.008

Habil M, Massey DD, Taneja A (2015) Exposure from particle and ionic contamination to children in schools of India. Atmos Pollut Res 6:719–725. https://doi.org/10.5094/APR.2015.080

Chithra VS, Nagendra SMS (2014) Characterizing and predicting coarse and fine particulates in classrooms located close to an urban roadway. J Air Waste Manage Assoc 64:945–956. https://doi.org/10.1080/10962247.2014.894483

Chithra VS, Shiva Nagendra SM (2012) Indoor air quality investigations in a naturally ventilated school building located close to an urban roadway in Chennai, India. Build Environ 54:159–167. https://doi.org/10.1016/j.buildenv.2012.01.016

Goel A, Ola D, Veetil AV (2019) Burden of disease for workers attributable to exposure through inhalation of PPAHs in RSPM from cooking fumes. Environ Sci Pollut Res. https://doi.org/10.1007/s11356-019-04242-x

Cong XC, Zhao JJ, Jing Z et al (2018) Indoor particle dynamics in a school office: determination of particle concentrations, deposition rates and penetration factors under naturally ventilated conditions. Environ Geochem Health 40:2511–2524. https://doi.org/10.1007/s10653-018-0116-x

Kumar A, Srivastava D, Agrawal M, Goel A (2014) Snapshot of PM loads evaluated at major road and railway intersections in an urban locality. Int J Environ Prot 4:23

McGarry P, Morawska L, He C et al (2011) Exposure to particles from laser printers operating within office workplaces. Environ Sci Technol 45:6444–6452

Bozkurt Z, Doğan G, Arslanbaş D et al (2015) Determination of the personal, indoor and outdoor exposure levels of inorganic gaseous pollutants in different microenvironments in an industrial city. Environ Monit Assess 187:590. https://doi.org/10.1007/s10661-015-4816-8

Habil M, Massey DD, Taneja A (2013) Exposure of children studying in schools of India to PM levels and metal contamination: sources and their identification. Air Qual Atmos Health 6:575–587. https://doi.org/10.1007/s11869-013-0201-3

Rohra H, Tiwari R, Khare P, Taneja A (2018) Indoor-outdoor association of particulate matter and bounded elemental composition within coarse, quasi-accumulation and quasi-ultrafine ranges in residential areas of northern India. Sci Total Environ 631–632:1383–1397. https://doi.org/10.1016/j.scitotenv.2018.03.095

Status of Carbonaceous Aerosol at Indoor Environment of a Cafeteria in Delhi, India—A Case Study

Papiya Mandal, M. Sri Nagesh and Anubha Mandal

Abstract The present study investigated the carbonaceous aerosol with respect to organic carbon (OC), elemental carbon (EC) and total carbon (TC) in particulate matter (PM_{10}) in indoor environment of cafeteria located at Netaji Subhash Place, one of the hotspot locations for pollution in northwest district of Delhi during 2014–15 winter season. The collections of samples were carried out during the period of three months (December 2014 to February 2015). PM_{10} samples were collected by APM 800 samplers (Envirotech Pvt. Ltd., India) on Whatman 37 mm microfiber quartz filter papers for 2–3 hourly bases in the dining area of food court. The flow rate varied from 2.4 lpm to 3 lpm during the period of collection of samples. Indoor PM_{10} concentrations varied from 1830 to 3212 $\mu g/m^3$. The concentration of OC, EC and TC in PM_{10} size of particulate matter varied from 54 to 318, 11 to 71 and 70 to 364 $\mu g/m^3$, respectively. The percentage contribution of OC and EC in TC were varied from 80 to 90% and 10 to 20%, respectively. The percentage contribution of TC in PM_{10} varied from 10 to 20%, respectively. The concentration of PM_{10} at indoor environment of cafeteria was alarmingly high as compared to National Ambient Air Quality Standard (NAAQS), 2009. The present study revealed that concentrations of PM_{10}, OC and EC at indoor environment of cafeteria were influenced by indoor and outdoor air pollution, meteorological parameters and guest count.

Keywords Aerosol · Indoor air pollution · Particulate matter · Delhi

1 Introduction

Indoor air quality (IAQ) is still relatively unexposed as compared to outdoor air quality. People spend their major time at indoor environment. Cafeteria may be considered as public indoor environment [1] where people love to spend their time

P. Mandal (✉)
CSIR—NEERI, Zonal Centre, Delhi, India
e-mail: p_mandal@neeri.res.in

M. Sri Nagesh · A. Mandal
Department of Environmental Engineering, Delhi Technological University, Delhi, India

A. Sharma et al. (eds.), *Indoor Environmental Quality*,
Lecture Notes in Civil Engineering 60,
https://doi.org/10.1007/978-981-15-1334-3_2

with favorite food, happy moments and discussions. Healthy IAQ of cafeteria not only provide the enjoyment to the public and workers of cafeteria but also protect them from the exposing to harmful air pollutants of the outdoor environment. A few studies have reported that IAQ is worst as compared to outdoor air quality. But people cannot understand because they spent short duration inside the cafeteria. At present scenario, air pollution at indoor environment is a grave problem in most of the mega cities in India and is receiving a lot of interest to the researchers. It is linked with socio-economic status of the public. Mostly, scientists are concentrated to atmospheric aerosols in present-day scenario due to growing anthropogenic activities as well as in terms of their effects on human health and climate change. Carbonaceous aerosols are one of the dominating contributors to atmospheric aerosols. They are the single largest absorber of solar radiation, and their heterogeneous reactions change the dynamics of the atmospheric boundary layer which reduces visibility and also create health hazard. Carbonaceous aerosol in terms of total carbon (TC) classified into two categories, i.e., organic carbon (OC) and elemental carbon (EC), which are the important constituents of PM_{10} (particulate matter 10 micrometers or less in diameter). OC is emitted from biogenic and anthropogenic sources, whereas EC is emitted only from anthropogenic sources.

Poor IAQ might be due to tiny particles (dust, pollen, animal fur, hair, etc.), harmful gases (cigar, fragrances, cooking odor, mosquito repeller, etc.), volatile compounds (floor cleaning substances, paint, glues, preservatives, etc.) and microorganisms (bacteria, fungus, virus, etc.) The major sources of poor IAQ might be due to combustion, furniture, building materials, outdoor air pollution and its penetration to indoor environment. IAQ also affected by building characteristics such inside space, orientation of the building, custom habit and tradition of the inhabitant, usage of equipment inside the building and ventilation system. The concentration of air pollutants at indoor environment also varies from location to location and season to season due to change in meteorological parameters (viz, *wind speed, wind direction, temperature, relative humidity, mixing height, etc.).* The common symptoms of indoor air pollution are headache, irritation of eyes, running nose, cough and cold, itchy throat, nausea, vomiting, etc. The potential health effect like respiratory infections, chronic bronchitis, development of cancers to multiple organs are due to outdoor air pollution or indoor air pollution or combination of exposure to both indoor–outdoor air pollution are still now on debate.

Despite arduous efforts put in research, the predominant sources contributing to indoor air pollution remained debatable. This is because, unlike other dining restaurants, cooking is carried out throughout the day in cafeteria. The varieties of food preparation use a wide variety of cooking oils as well as cooking fuels like LPG/PNG or coal. The heating of cooking oil and cooking fuels are the major source of indoor carbonaceous aerosol. It is evident from the study that inflated concentrations of particulate matter (PM_{10}) (particulate matter 10 micrometers or less in diameter), organic carbon (OC) and elemental carbon (EC) prevailed in the cafeteria. OC, a submicron particle [2] of combustion, and EC, an essential primary pollutant [3] emitted by incomplete combustion of fossil fuels [4], are the significant contributors to particulate matter [5]. Yearlong observation of concentration of OC and EC by various researchers revealed a trend in indoor air pollution which is particularly high during winter season and low during the monsoon season. People spending their maximum time at indoors (such as in offices, cafeterias and houses) are mostly exposed to greater concentrations of pollutants in urban areas [6]. There is the need for continuous monitoring and analysis of carbonaceous aerosols (OC, EC and TC) in size segregated size of particulate matter at indoor environment. Till date, no studies have been reported in India regarding the concentrations of indoor air pollutants with respect to PM_{10}, and its correlation with carbonaceous aerosols (OC, EC and TC). The effort was made to put-forth the scenario of PM_{10} and TC that comprised of OC and EC during the winter season at cafeteria located in Netaji Subhash Place, one of the hotspot locations at northwest district of Delhi, India.

2 Methodology

2.1 Sampling Location

Delhi, the capital of India, is considered to be one of the most polluted cities in the world by [7]. The inland position and the continental air prevalence influenced the semi-arid climatic condition in Delhi. The climate of Delhi varies from arid to semi-arid. Winter season is moderately cold and pre-monsoon is extremely hot with frequent dust storms. The annual rainfall in Delhi varies from 600 to 800 mm and maximum rainfall occurs during the monsoon season only. The temperature ranges between 1 °C and 48 °C during winter to summer season. Minimum temperatures and foggy conditions during winters trigger inversion condition, which leads to the accumulation of atmospheric air pollutants [8]. The air samples were collected from inside the cafeteria of Netaji Subhash Place at ground level during the period of December 2014 to February 2015. The sampling location of cafeteria at Netaji Subhash Place is shown in Fig. 1.

The instrument was placed at the juncture of dining and kitchen activities, within the premises of the cafeteria. The cafeteria is situated in hotspot location of West Delhi adjacent to major arterial busy traffic roads. During the air sampling, precautionary measures were taken not to expose the instrument directly to the outdoor environment. The cafeteria is located at the ground level. Flow rate of the monitoring instrument was maintained almost the same for the time period of monitoring. The sampling location inside the cafeteria is shown in Fig. 2.

3 Meteorological Condition

The data of meteorological parameters (such as temperature, wind speed, wind direction and relative humidity) were collected from the India Meteorological Department

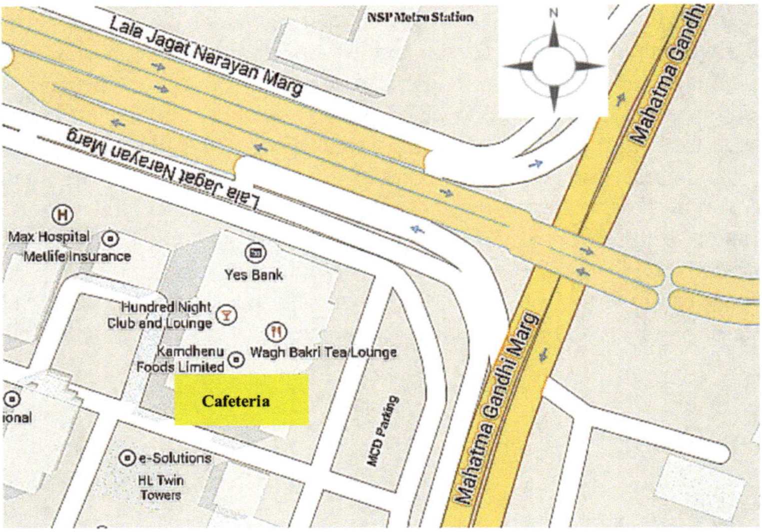

Fig. 1 Sampling location of cafeteria at Netaji Subhash Place (Google Map)

Fig. 2 The sampling location inside the cafeteria

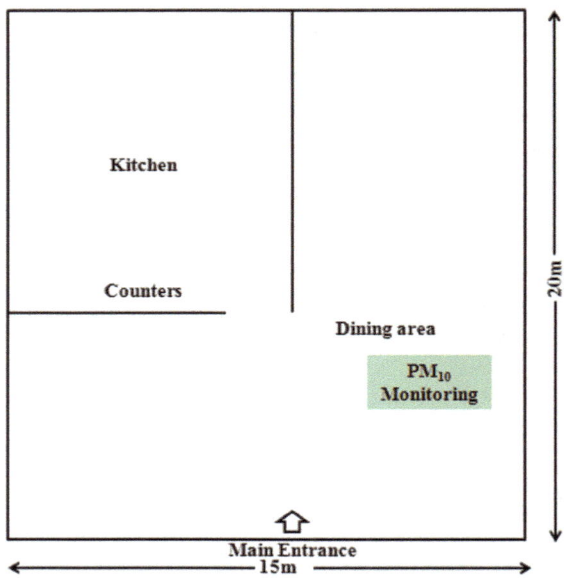

(IMD) and also from the Bhuvan Panchayat developed by NRSC–ISRO [9], respectively. The minimum and maximum temperatures of Delhi varied between 2 °C and 29 °C, respectively during the study period. The average temperature, relative humidity and wind speed of Delhi were 12.5 °C, 59.5%, and 4.00 knots, respectively.

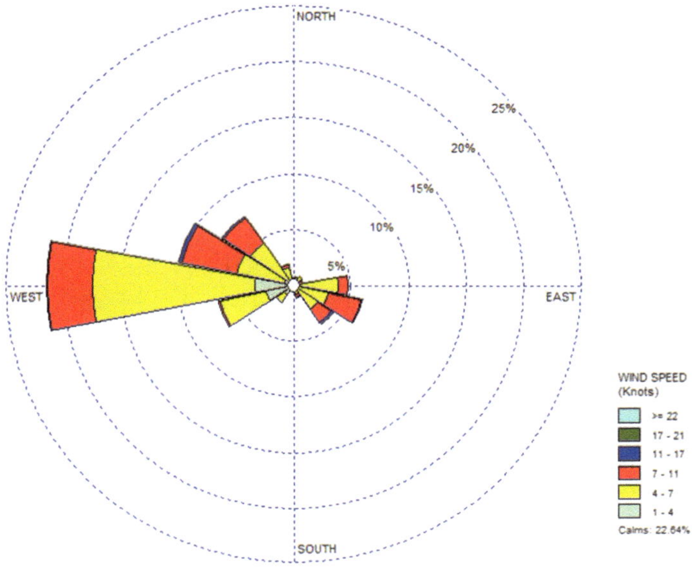

Fig. 3 Wind rose plot over Delhi (December 2014 to February 2015)

The prevailing wind directions were west and northwest directions during the study period. The wind rose plot over the study period is shown in Fig. 3.

4 Instrumentation (PM_{10} and Carbonaceous Aerosol)

PM_{10} samples were collected on Whatman microfiber quartz filter papers during the peak hours (1:00 p.m. to 4:00 p.m.) claimed by the cafeteria manager at a flow rate variation of 2.4 to 3.0 lpm using APM 800 sampler (make Envirotech Pvt. Ltd., Delhi, India). The filter papers were pre-baked in a muffle furnace at 550 °C for 6 h only to remove organic impurities. They were also pre-desiccated and post-desiccated for 24 h [10]. The concentrations of PM_{10} ($\mu g/m^3$) were measured by gravimetric method using Sartorius microbalance with accuracy till six decimals.

OC (composition of aliphatic, aromatic compounds, acids etc.) and EC (primary pollutant) analyses were carried through Carbon Analyzer by Interagency Monitoring of Protected Visual Environment (IMPROVE) and thermal optical reflectance (TOR) protocol (The DRI Model 2001). The basic principle of the analyzer is to oxidize OC and EC at different temperatures to get their respective fractions [11]. Analysis sequence of the IMPROVE protocol is OC, pyrolyzed carbon fraction (PY) and EC. OC is further divided into OC1, OC2, OC3 and OC4 at temperatures 140 °C, 280 °C, 480 °C and 580 °C, respectively; and EC is further divided into EC1, EC2 and EC3 at temperatures 580 °C, 740 °C and 840 °C, respectively and analyzed. Predefined OC by the IMPROVE protocol is: OC = OC1 + OC2 + OC3 + OC4 + PY. Similarly, EC is predefined as: EC = EC1 + EC2 + EC3 – PY and TC is predefined as: TC = OC + EC [12, 13].

5 Results and Discussion

The collected particulate matter (PM_{10}) samples during the period of December 2014 to February 2015 were analyzed for organic carbon (OC), elemental carbon (EC) and total carbon (TC) for both weekdays and weekends to understand the maximum concentrations of indoor air pollution at cafeteria during the winter season. No standards or guidelines were established till date in India to control the increasing PM_{10} concentration at indoor environment. In fact, no studies have been carried out in India regarding the concentrations of air pollutants with respect to PM_{10} and its correlation with carbonaceous aerosols (OC, EC and TC). The statistics of concentrations of PM_{10}, OC, EC and TC at cafeteria, during the study period (December 2014 to February 2015) is shown in Table 1.

The concentrations of PM_{10} in weekdays and weekends varied from 2347 to 2976 and 1830 to 3212 $\mu g/m^3$ with an average concentration of 2711 ± 161 (weekdays) and 2708 ± 494 (weekends) in $\mu g/m^3$, respectively. The concentrations of PM_{10} at indoor environment of cafeteria were alarmingly high and beyond the permissible limit (100 $\mu g/m^3$) of National Ambient Air Quality Standards (NAAQS) formulated by Central Pollution Control Board, 2009 [14]. The highest concentrations of PM_{10} in both weekdays and weekends were observed in the month of December 2014 due to celebration of enjoyable moments of Christmas and New Year. Festive season, people prefer to celebrate inside the cafeteria, which indicated visiting of maximum guest count in the cafeteria. The maximum visiting guest count of the cafeteria indicated maximum concentration of particulate matter (PM_{10}) might be due to maximum

Table 1 Statistics of PM_{10}, OC, EC and TC at indoor environment of cafeteria in Delhi, India

	PM_{10} ($\mu g/m^3$)	OC ($\mu g/m^3$)	EC ($\mu g/m^3$)	TC ($\mu g/m^3$)	OC/EC
Weekdays					
Min	2347	99	11	116	3
Max	2976	318	61	364	12
Average	2711	174	29	203	7
Std. Dev.	161	76	17	88	3
Weekends					
Min	1830	54	11	70	3
Max	3212	214	71	280	9
Average	2708	141	31	171	5
Std. Dev.	494	60	19	74	2

cooking activities and maximum penetration of outdoor air pollutants inside the cafeteria. The unfavorable meteorological conditions during the winter season such as stability of atmosphere, slow dispersion and low-average mixing height were the additional parameters which increased the concentration of PM_{10} at the indoor environment of cafeteria.

The concentrations of OC, EC and TC in PM_{10} in weekdays varied from 99 to 318 and 11 to 61 and 116 to 364 $\mu g/m^3$ with an average concentration of 174 ± 76, 29 ± 17 and 203 ± 88in $\mu g/m^3$, respectively. The concentrations of OC, EC and TC in PM_{10} in weekends varied from 54 to 214 and 11 to 71 and 70 to 280 $\mu g/m^3$ with an average concentration of 141 ± 60, 31 ± 19 and 171 ± 74 in $\mu g/m^3$, respectively. There was not much variation of concentration of EC in both weekdays and weekends as it is the primary pollutant. In general, EC increased in the air environment due to emission of soot particles from usage of biofuels, vehicular and industrial emission. The concentration of OC concentration varied in large extent might be due to gas to particle conversation and emission from anthropogenic sources like tobacco smoking, usage of, emission of soot particles from usage of biofuels, vehicular and industrial emission. Weekdays OC concentrations were higher as compared to weekends might be due to movement of less vehicles in weekends as compared to weekdays. The percentage contribution of OC and EC in TC was varied season to season. It was observed that concentration of OC and EC in TC in PM_{10} in an average varied from 80 to 90% and 10 to 20%, respectively. The concentration of TC is one of the major constituents in PM_{10} and at indoor environment the percentage contribution of TC in PM_{10} varied from 10 to 20%, respectively. The average OC and EC concentration inside the cafeteria, varied from 3 to 12. This indicated the burning of biomass was the major source [15]. In general, the ratio of OC and EC greater than two indicated the presence of secondary organic carbon (SOC), but mostly ignored due to prevailing low temperatures, especially during the winter season, which may contribute to meager formations of SOC [16] in TC in the indoor environment.

6 Conclusions

The present IAQ study with respect to PM_{10} associated OC and EC during winter season at cafeteria revealed that poor IAQ. PM_{10} concentration was abnormally high as compared to NAAQS in Delhi, the capital of India. People inside the cafeteria are exposed to high concentrations of PM_{10} bound OC and EC for short duration of time which may not be realized. The most effective way to improve IAQ is to reduce the sources of chemical emissions which may include usage of cleaning substances, furnishings, furniture, flooring, paint, textiles building materials, etc. Tobacco smoking, cooking activities, fuel usages, exchange of indoor and outdoor air pollutants Meteorological conditions are also the governing factors to increase indoor concentration of PM_{10} and carbonaceous aerosols. In the fully centralized air conditioning cafeterias, periodical cleaning of air conditioning systems is essential. The frequent cleaning of floor of the cafeterias may reduce the emission of resuspension of floor dust. The provision to place indoor dust-capturing plants will definitely reduce the concentration of PM_{10} as well as PM_{10} bound OC and EC of the cafeterias.

Acknowledgements The authors are thankful to the Director, CSIR NEERI to grant them the permission to conduct study at the Zonal Centre in Delhi. The authors are thankful to Environmental Engineering Department, DTU to allow and to publish the research work.

References

1. Lee SC, Li WM, Chan LY (2001) Indoor air quality at restaurants with different styles of cooking in metropolitan Hong Kong. Sci Total Environ 279:181–193
2. Castro LM, Pio CA, Harrison RM, Smith DJT (1999) Carbonaceous aerosol in urban and rural European atmospheres: estimation of secondary organic carbon Concentrations. Atmos Environ 33:2771–2781
3. Ram K, Sarin MM (2008) Atmospheric abundances of primary and secondary carbonaceous species at two high-altitude sites in India: sources and temporal variability. Atmos Environ 42:6785–6796. https://doi.org/10.1016/j.atmosenv.2008.05.031
4. Salam A, Bauer H (2003) Aerosol chemical characteristics of a mega-city in Southeast Asia (Dhaka–Bangladesh). Atmos Environ 37:2517–2528. https://doi.org/10.1016/S1352-2310(03)00135-3
5. Ye D, Zhao Q (2007) Characteristics of elemental Carbon and organic carbon in PM_{10} during spring and autumn in Chongquig, China. China Particuol 5:255–260. https://doi.org/10.1016/j.cpart.2007.03.009
6. Ho KF, Cao JJ (2004) Indoor/outdoor relationships of organic carbon (OC) and elemental carbon (EC) in $PM_{2.5}$ in roadside environment of Hong Kong. Atmos Environ 38:6327–6335. https://doi.org/10.1016/j.atmosenv.2004.08.007
7. World Health Organization (2014) Copy of aap_pm_database_may2014.xls. http://www.who.int/phe/health_topics/outdoorair/databases/cities/en/. Accessed 15 Feb 2016
8. Tiwari S, Srivastava AK (2013) Diurnal and seasonal variations of black carbon and $PM_{2.5}$ over New Delhi, India: Influence of Meteorology. Atmos Res 125–126:50–62. https://doi.org/10.1016/j.atmosres.2013.01.011
9. BhuvanPanchayats—Space based information support for decentralized planning (SIS—DP) http://bhuvan-panchayat.nrsc.gov.in. Accessed 18 Dec 2015
10. Mandal P, Saud T (2013) High seasonal variation of atmospheric C and particle concentrations in Delhi, India. Environ Chem Lett. https://doi.org/10.1007/s10311-013-0438-y
11. Mandal P, Sarkar R (2014) Seasonal variation and sources of aerosol pollution in Delhi, India. Environ Chem Lett. https://doi.org/10.1007/s10311-014-0479-x
12. Park SS, Kim YJ, Kang CH (2002) Atmospheric polycyclic aromatic hydrocarbons in Seoul, Korea. Atmos Environ 36:2917–2924
13. Chow JC, Watson JG (2012) The IMPROVE_A temperature protocol for thermal/optical carbon analysis: maintaining consistency with a long-term database. J Air Waste Manag Assoc 57(9):1014–1023. https://doi.org/10.3155/1047-3289.57.9.1014
14. Central Pollution Control Board CPCB (2009). National ambient air quality standards in India. http://cpcb.nic.in/uplo/Publication/Publication_514_airqualitystatus2009.pdf

15. Safai PD, Raju MP (2014) Characterization of carbonaceous aerosols over the urban tropical location and a new approach to evaluate their climatic importance. Atmos Environ 92:493–500. https://doi.org/10.1016/j.atmosenv.2014.04.055
16. Lee S, Wang Y (2010) Assessment of secondary organic carbon in the southeastern United States: a review. J Air Waste Manag Assoc 60:1282–1292. https://doi.org/10.3155/1047-3289.60.11.1282

Assessment of Indoor Fine and Ultra-Fine Particulate Matter in a Research Laboratory

Amit K. Mishra, P. Mishra, Sunil Gulia and S. K. Goyal

Abstract Indoor air quality (IAQ) has drawn the attention of all the scientific community around the globe as it ranked one of the top five risks to public health throughout the world. People spend most of their time in indoor environments, be it in their home or workplace, without knowing that they are inhaling substantially high concentrations of different indoor air pollutants (IAPs). In developing countries, IAP concentrations are generally found high due to poor ventilation and numerous indoor sources. Poor IAQ can adversely obstruct the mental, physical and social ability of a person, which can affect the working efficiency and result into loss in overall productivity. Along with other IAPs, high level of PM in indoor environments is one of the major concerns. The present study is an attempt to assess the exposure levels of PM_{10}, $PM_{2.5}$ and PM_1 in one of the research laboratories, located in an industrial area of Delhi city. The monitoring is carried out at different indoor environments of the building. The preliminary results indicate that average concentration of PM_{10}, $PM_{2.5}$ and $PM_{1.0}$ are highest in chemical laboratory, i.e., $114 \pm 25\ \mu g/m^3$, $58 \pm 10\ \mu g/m^3$, $33 \pm 5\ \mu g/m^3$, respectively and lowest at un-disturbed area, i.e., $42 \pm 4\ \mu g/m^3$, $33 \pm 2\ \mu g/m^3$, $22 \pm 2\ \mu g/m^3$, respectively. Ratio of $PM_{2.5}/PM_{10}$ and $PM_{1.0}/PM_{2.5}$ are found higher at un-disturbed area, i.e., 0.79 and 0.75, respectively as compared to other areas. It indicates that fine and ultra-fine particles travel more from outdoor compared to larger particle and suspend for longer time in the environment. Further, the study also discusses the possible measures to control the indoor air pollution and prioritize the indoor air-purifying plants based on their efficiency.

A. K. Mishra · P. Mishra · S. Gulia (✉) · S. K. Goyal
CSIR-National Environmental Engineering Research Institute (NEERI), Delhi Zonal Centre, Naraina, New Delhi 110028, India
e-mail: sunilevs@gmail.com

A. Sharma et al. (eds.), *Indoor Environmental Quality*,
Lecture Notes in Civil Engineering 60,
https://doi.org/10.1007/978-981-15-1334-3_3

Keywords Indoor air quality · Particulate matter · IAQ purifying plants · Ventilation

1 Introduction

Indoor air pollution is one of the major causes of global burden diseases in the world and ranked as one of the top five risks to public health. Most of the cases, indoor air pollution is more harmful than the outdoor air pollution, because confined areas enable potential pollutants to build up extra than open spaces. Most of the people spend their maximum time in indoor and exposed higher compared to outdoors. On average 80% of time people spend indoors [1, 2]. Yu et al. [3] reported that unhealthy indoor air quality is the main contributor of "sick building syndrome" (SBS). High exposure in poor indoor pollution causes allergies, respiratory dysfunction, nasal irritations headache and fatigue to building occupants [3]. Additionally, exposure in polluted indoor workplace can reduce the work performance, output and the wellbeing of building occupants [4].

IAQ of any building is related to a different range of physical, chemical and biological factors like emission rate of chemicals and dust particle, the frequency of indoor air to exchange with ambient air, atmospheric circulation efficiency inside the building. The major indoor air pollution sources are combustion, building material, indoor activities and penetration of outdoor air pollution. The incomplete combustion of fuel emits particulate matters, hydrocarbons and other gaseous pollutants such as carbon monoxide, nitrogen oxides, sulfur dioxide. Some of the sources are associated with (i) indoor activities as housekeeping and maintenance like cleansers, disinfectants, air fresheners, mats and lubricants; (ii) occupant-related sources are tobacco product, office equipment (computers, printers and copiers), cooking stoves/microwaves, paper products and dirt/pollens; and (iii) building materials, i.e., plywood/compressed wood, wall panels, construction adhesives, carpets, tiles and also heating, ventilation and air conditioning systems like boilers, furnaces, generators and stoves.

Meng et al. [5] stated that ambient air can contribute about 25–65% of indoor air pollution [5]. In one of the studies, it was found that penetration of outdoor air pollution is one of the major sources of indoor air pollution in a typical building [6]. They reported that indoor/outdoor (I/O) ratio of any pollutant significantly affects outside meteorological factors and air exchange rate of the building [7, 8]. Researcher also found that in the absence of intense indoor and ambient sources plays an important role for high pollution level in indoor greater outdoor concentrations rather than indoor values [9, 10].

High level of fine particulate matter in the indoor environment mainly due to penetration from outdoor PM is one of the common sources in the naturally ventilated building. This problem is more complex in city like Delhi where ambient PM level is very high and exceeding the specified standards. PM is a composition of very small particles and composed of various chemicals such as heavy metals, organic and inorganic carbon, secondary ions and acids [11]. The toxicity of the PM varies based on its composition and types of sources. Chitra and Nagendra [12] have done a study in a naturally ventilated school in Chennai and found that 24-h average suspended particulate matter (SPM), PM_{10}, $PM_{2.5}$ and $PM_{1.0}$ concentrations were 168.64 $\mu g/m^3$, 135.88 $\mu g/m^3$, 42.95 $\mu g/m^3$ and 25.89 $\mu g/m^3$, respectively. Similarly, Singh et al. [13] have also done indoor air quality assessment in selected schools of Delhi-NCR and concluded that the average $PM_{2.5}$ concentrations in both air-conditioned and naturally aired school buildings were much above the levels suggested in National Ambient Air Quality Standards, India. Indoor air quality standards are not yet defined in India that is why compared with ambient standards. It seems that indoor PM level is found higher in Indian cities and need to be reduced for better indoor air quality.

Numerous techniques are available for removal of indoor air pollutants. These are mechanically removal of pollutants through air purifier; mechanically ventilation with high air exchange rate, removal of VOCs from high adsorbing building materials and planting air-purifying plants. Improvement of indoor air quality through plants is one of the efficient and economical methods [14–18]. Indoor air-purifying plants have many benefits. Plants can improve air quality through several mechanisms as they can use carbon dioxide and give oxygen during photosynthesis. By the process of transpiration, they can increase the humidity through very small leaf pores, and also they can absorb pollutants on the external surface of leaves and the plant root–soil system. Gawronska and Bakera [14] have found that spider plants help to accumulate PM. Troy and Zavattaro [15] were tested Chlorophytum comosum (spider plant) and Epipremnum aureum (pothos) for removal of pollutant from indoors and found that each species could significantly reduce PM.

The present study is mainly focusing on the assessment of different size PM in a research laboratory located in one of the industrial area of Delhi city, India. Further, the importance of indoor air pollution and related health issues were discussed along with possible solution through air-purifying plants.

2 Methodology

2.1 Site Description

The study is conducted in a research laboratory located in the Naraina Industrial area where various activities carried out in the surrounding area. A research laboratory is a workplace where anyone can find different work environments such as typical administrative office, scientific staff office, trainee student room and chemical laboratory area. The major sources outside the office area are commercial activities and vehicular movement. A slum area also presents near to the study site where domestic emission and open burning of biomass are common practices which also generates huge amount of PM.

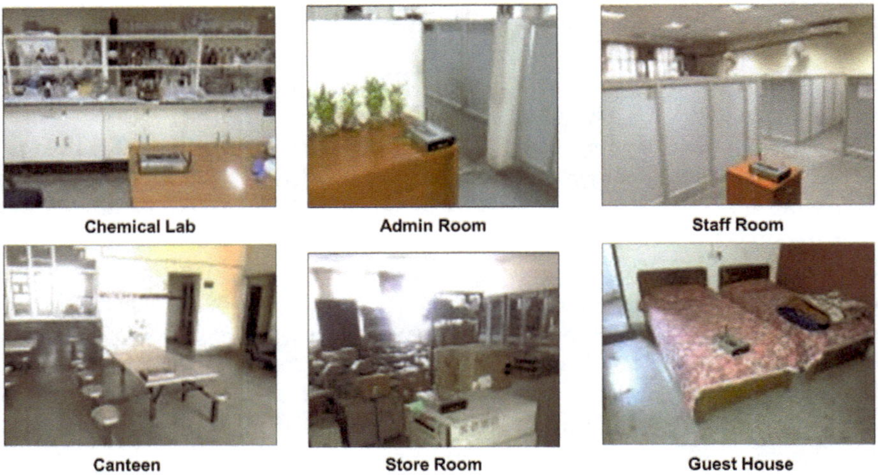

Fig. 1 Indoor air monitoring locations

2.2 *Indoor PM Monitoring*

Indoor air quality monitoring was carried out three days in the last week of August 2018 (August 23–25, 2018). The PM_{10}, $PM_{2.5}$ and $PM_{1.0}$ measurements were made in each of the selected indoor environment at a sampling frequency of 1 min using calibrated aerosol dust monitor (make GRIMM, R-11 model). The monitor performs on the principle of light scattering by sucking air with multiple size particle at a flow rate of 1.2 lit/min and passed through a laser beam. This is capable of measuring particle mass concentrations in the range of 1-6500 $\mu g/m^3$. A 31-channel pulse height analyzer for size classification detects the scattering signals in the size range of 0.3–25 μm. The monitoring is carried out at chemical laboratory, administration office, scientific/technical staff room, canteen area, guesthouse and storeroom (un-disturbed area). Figure 1 shows the photographs of different office area where monitoring performed. The monitoring is carried out for three consecutive days at the same time in the respective area, during working hours only.

3 Results and Discussion

The monitored data of PM_{10}, $PM_{2.5}$ and $PM_{1.0}$ were analyzed statistically and compared relatively. Tables 1 and 2 show 15 min average concentrations of PM_{10}, $PM_{2.5}$ and $PM_{1.0}$ for all the six indoor air environment.

The preliminary results indicate that 15 min average concentration of PM_{10}, $PM_{2.5}$ and $PM_{1.0}$ were found maximum in the chemical laboratory, i.e., 114 ± 25 $\mu g/m^3$, 58 ± 10 $\mu g/m^3$, 33 ± 5 $\mu g/m^3$, respectively and minimum at storeroom (un-disturbed

Table 1 Average PM concentration in different indoor chambers

Location	PM_{10} ($\mu g/m^3$)		$PM_{2.5}$ ($\mu g/m^3$)		PM_1 ($\mu g/m^3$)	
	Avg.	SD	Avg.	SD	Avg.	SD
Chemical laboratory	114	25	58	10	33	5
Admin room	56	11	32	5	23	3
Staff room	50	7	33	4	22	3
Store room	42	4	33	2	25	2
Canteen	104	28	57	9	34	3
Guest house	80	23	46	3	29	5

Avg.—average, *SD*—standard deviation

Table 2 Average ratio of PM $_{2.5}$/PM $_{10}$, PM $_{1.0}$/PM $_{10}$ and PM $_{1.0}$/PM $_{2.5}$

Location	$PM_{2.5}/PM_{10}$	$PM_{1.0}/PM_{10}$	$PM_{1.0}/PM_{2.5}$
Chemical laboratory	0.51	0.29	0.56
Admin room	0.58	0.41	0.71
Staff room	0.65	0.45	0.69
Store room	0.79	0.59	0.75
Canteen	0.55	0.33	0.59
Guest house	0.57	0.36	0.63

area), i.e., 42 ± 4 $\mu g/m^3$, 33 ± 2 $\mu g/m^3$, 22 ± 2 $\mu g/m^3$, respectively. Further, the ration of $PM_{2.5}/PM_{10}$ and $PM_{1.0}/PM_{2.5}$ were calculated to evaluate the proportion of different size particles in the indoor air. It is observed that ratio of $PM_{2.5}/PM_{10}$ and $PM_1/PM_{2.5}$ were found higher at store area, i.e., 0.79 and 0.75, respectively as compared to other office environments where these values were found less than 0.59 and 0.70, respectively. The higher ration of $PM_{2.5}/PM_{10}$, $PM_{1.0}/PM_{10}$ and $PM_{1.0}/PM_{2.5}$ in storeroom which opens rarely were indicates that fine and ultra-fine particles travel more from outdoor through leakages present in the door/window compared to larger particle. The proportion of $PM_{1.0}$ in PM_{10} was found in the range of 0.29 (chemical laboratory)–0.59 (storeroom). The larger particles are generally resuspended in the indoor environment through movement and other activities which is indicated in the monitored level. Figure 2 shows the average concentration of PM_{10}, $PM_{2.5}$ and $PM_{1.0}$ for the study period at different workplace which indicates a high variation in the PM_{10} and $PM_{2.5}$ level. However, there is not much variation in $PM_{1.0}$ level and found more or less similar at all place.

In the past, researchers also reported similar pattern of PM concentration in the indoor environment. In one of the studies, assessed IAQ in an office building and found average PM_{10} and $PM_{2.5}$ concentrations of 21.2 $\mu g/m^3$ and 15.5 $\mu g/m^3$, respectively [19]. Goyal and Kumar [20] found that I/O ratio for PM_{10}, $PM_{2.5}$ and $PM_{1.0}$ varied from 0.37–3.1, 0.2–3.2 and 0.17–2.9 $\mu g/m^3$, respectively, at commercial building in Delhi city. In a school building in Delhi, the I/O ratio of PM_{10}, $PM_{2.5}$ and $PM_{1.0}$

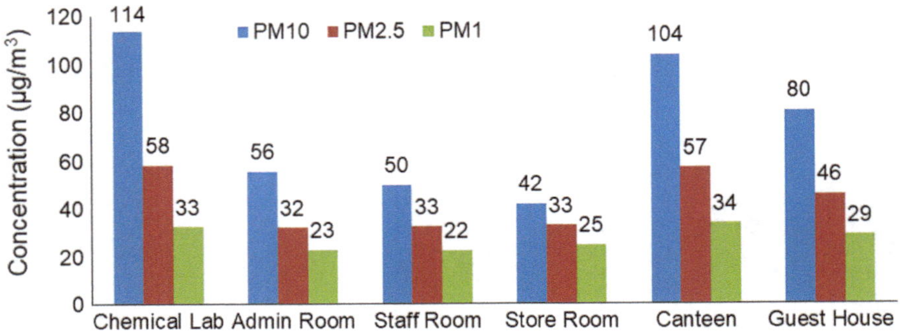

Fig. 2 Graphical representation of PM concentration in different indoor chambers

were found to be 2.52 ± 2.71, 1.44 ± 0.67 and 0.97 ± 0.18, respectively, which indicate higher value because of occupant's activities inside the building due to research scholar movements and lower value due to outdoor air pollution [12].

The aerosol dust monitor measured particles having size range from 0.25 μm to 32.0 μm. The proportion of concentration of different particle size is plotted and compared between different indoor work environments (Fig. 3). It is observed that pattern of storeroom areas (non-active area) are totally different than other indoor environments of the office which are generally active area. In storeroom, maximum portion is due to finer particle, i.e., 69% by $PM_{2.5}$ and 95% by PM_{10} when compared to other workplaces where higher portion is due to larger size particles. The concentration pattern up to $PM_{2.5}$ is almost similar in all indoor environments except storeroom. However, pattern in guest hour and canteen are more or less similar and staff room, admin area and chemical laboratory are similar. Further, this correlates

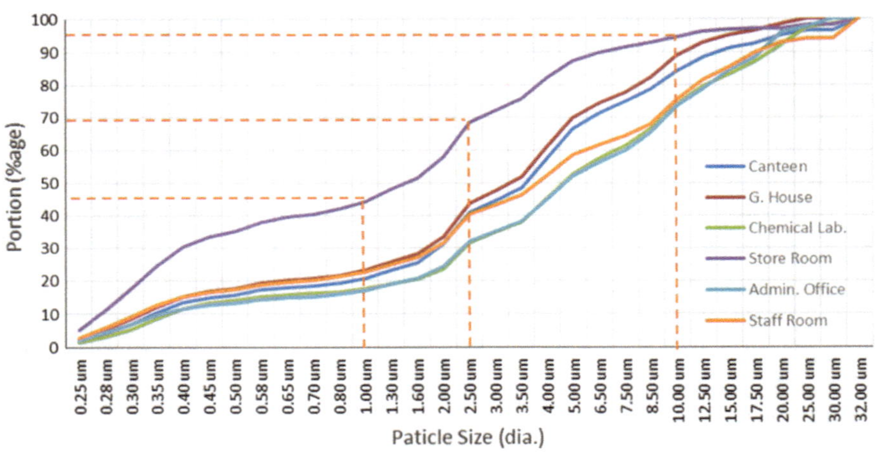

Fig. 3 Proportion of different size particles in different work environment

well with the type of activity level. It is indicated that particle having size between 2.5 and 10 μm are highly influenced by indoor activity and resuspended due to movement and other cleaning activities.

4 Conclusion

Indoor air pollution is one of the major causes of global burden diseases and reduces the work performance in the office. Indoor PM_{10}, $PM_{2.5}$ and $PM_{1.0}$ were monitored at different work environment of a research laboratory in Delhi city. The results indicate that average concentration of PM_{10}, $PM_{2.5}$ and $PM_{1.0}$ were found maximum in the chemical laboratory, i.e., 114 ± 25 μg/m^3, 58 ± 10 μg/m^3, 33 ± 5 μg/m^3, respectively and minimum at storeroom (un-disturbed area), i.e., 42 ± 4 μg/m^3, 33 ± 2 μg/m^3, 22 ± 2 μg/m^3, respectively. High variations were observed in the level of PM_{10} and $PM_{2.5}$ between different office areas, however, not much variation observed in the $PM_{1.0}$ level. This indicates that resuspension of PM is one of the major sources of Indoor PM_{10} concentrations. It is also observed that portion of ultra and fine particles in PM concentration were observed in the indoor environment with minimum disturbance compared to office area where activities are more. High concentration of indoor PM should be removed through indoor air-purifying plants as they provide cleaner and healthier air to us. They can also absorb pollutant on their outer surface. Leaf size, structure, the thicker layer of waxes, pubescence and surface roughness usually correspond to a higher absorption of pollutants from both indoor and outdoor polluted air [15–17].

References

1. World Health Organization (2010) WHO guidelines for indoor air quality: selected pollutants. WHO, Geneva
2. Bluyssen PM, Roda C, Mandin C, Fossati S, Carrer P, de Kluizenaar Y, Mihucz VG, de Oliveira Fernandes E, Bartzis J (2016) Self-reported health and comfort in 'modern' office buildings: first results from the European OFFICAIR study. Indoor Air 26:298–317
3. Yu BF, Hu ZB, Liu M, Yang HL, Kong QX, Liu YH (2009) Review of research on air conditioning systems and indoor air quality control for human health. Int J Refrig 32:3–20
4. Wood R, Burchet M, Alquezar R, Orwell R, Tarran J, Torpy F (2006) The potted plant microcosm substantially reduces indoor air VOC pollution I. Office field study. Water Air Soil Poll. 175:163–180
5. Meng QY, Turpin BJ, Korn L, Weisel CP, Morandi M, Colome S, Zhang J, Stock T, Spektor D, Winer A, Zhang L, Lee JH, Giovanetti R, Cui W, Kwon J, Alimokhtari S, Shendell D, Jones J, Farrar C, Maberti S (2004) Influence of ambient (outdoor) sources on residential indoor and personal $PM_{2.5}$ concentrations: analyses of RIOPA data. J Expo Anal Environ Epidemiol 15:17–28
6. Katiyar V, Khare M (2008) Indoor Air quality—an emerging environmental challenge. In: CES Newsletter Enviroscan

7. Lachenmyer C (2000) Urban measurements of outdoor-indoor $PM_{2.5}$ concentrations and personal exposure in the deep South. Part I. Pilot study of mass concentrations for nonsmoking subjects. Aerosol Sci Technol 32:34–51
8. Fundamentals of Indoor Air Quality in Buildings, EPA. https://www.epa.gov/indoor-air-quality-iaq/fundamentals-indoor-air-quality-buildings. Accessed on 18 Mar 2019
9. Hoek G, Kos G, Harrison R, de Hartog J, Meliefste K, Ten Brink H, Katsouyanni K, Karakatsani A, Lianou M, Kotronarou A (2008) Indoor–outdoor relationships of particle number and mass in four European cities. Atmos Environ 42:156–169
10. Monn C (2001) Exposure assessment of air pollutants: A review on spatial heterogeneity and indoor/outdoor/personal exposure to suspended particulate matter, nitrogen dioxide and ozone. Atmos Environ 35:1–32
11. Anonymous (2011) Particulate matter air and radiation US EPA US environmental protection agency. http://www.epa.gov/pm/. Accessed on 22 July 2011
12. Chithra VS, Nagendra SMS (2018) Source identification of indoor particulate matter and health risk assessment in school children. https://doi.org/10.1061/(ASCE)HZ.2153-5515.0000390
13. Singh P, Arora R, Goyal R (2017) Indoor air quality assessment in selected schools of Delhi-NCR, India. Int J Appl Home Sci 4(5&6):389–394
14. Gawronska H, Bakera B (2015) Phytoremediation of particulate matter from indoor air by Chlorophytum comosum L. plants. Air Qual Atmos Health 8:265–272
15. Troy F, Zavattaro M (2018) Bench study of green-wall plants for indoor air pollution reduction. J Living Archit 5(1):1–15
16. Khean CC, Yih TY (2018) A review of the Effective study of Botanical components that improves air filtration. In: International Conference on Earth and Environmental Science, vol 167
17. Wei X (2017) Phylloremediation of air pollutants: exploiting the potential of plant leaves and leaf-associated microbes. Front Plant Sci 8:1318
18. Saebo A, Popek R, Nawrot B, Hanslin HM, Gawronska H, Gawronski SW (2012) Plant species differences in particulate matter accumulation on leaf surfaces. Sci Total Environ 15(427–428):347–354 (Epub)
19. Dikaia S, Thomas M, Eman S, Eleni F, Hala H, Vasiliki T, Sotirios K, Aikaterini S, Christos V, Konstantinos K (2017) Chemical characterization of indoor and outdoor particulate matter ($PM_{2.5}$, PM_{10}) in Doha. Qatar. Aerosol Air Qual Res 17:1156–1168. https://doi.org/10.4209/aaqr.2016.05.0198
20. Goyal R, Kumar P (2013) Indoor–outdoor concentrations of particulate matter in nine microenvironments of a mix-use commercial building in megacity Delhi. Air Qual Atmos Health 6:747–757. https://doi.org/10.1007/s11869-013-0212-0

Air Pollution in Rural Households Due to Solid Biomass Fuel Use and Its Health Impacts

Khaiwal Ravindra, Maninder Kaur-Sidhu and Suman Mor

Abstract Air pollution has become a major environmental health risk factor in lower- and middle-income countries. There are several policies and regulations to control outdoor air pollution, but household air pollution is ignored. Children, elderly, and women, who spend most of their time in the indoor environment are at the higher risks of exposure due to the uses of solid biomass fuels for household energy. The exposure to toxic household air pollutants from incomplete combustion of solid biomass fuels has been linked with chronic obstructive pulmonary disease, acute lower respiratory infections, lung cancer, stroke, ischaemic heart disease, etc. However, evidence from countries where household air pollution has the highest adverse impact, lacks to convince the policymakers. Hence, there is a need to focus on personal exposure assessment to establish the causal effect relationship. Recently, several governments in Asia, including India, have taken initiatives to curb the use of solid biomass fuels. This provides an opportunity to examine the benefits of better health and the environment.

Keywords Household air pollution · Biomass fuels · Noncommunicable diseases

1 Air Pollution and Health

Pollution is the largest environmental cause of disease with a significant contribution from air pollution. The vulnerable population living in developing countries are highly exposed to household and ambient air pollutants, as depicted in Fig. 1 [1]. Figure 2 shows the traditional cookstove, which uses solid biomass such as cow dung cake, wood, or crop residue as fuel. Approximately, three billion people globally are dependent on inefficient, polluting fuels, which contribute to 85% of

K. Ravindra (✉) · M. Kaur-Sidhu
Department of Community Medicine and School of Public health, Post Graduate Institute of Medical Education and Research (PGIMER), Chandigarh 160012, India
e-mail: khaiwal@yahoo.com

S. Mor
Department of Environmental Studies, Panjab University, Chandigarh, India

A. Sharma et al. (eds.), *Indoor Environmental Quality*,
Lecture Notes in Civil Engineering 60,
https://doi.org/10.1007/978-981-15-1334-3_4

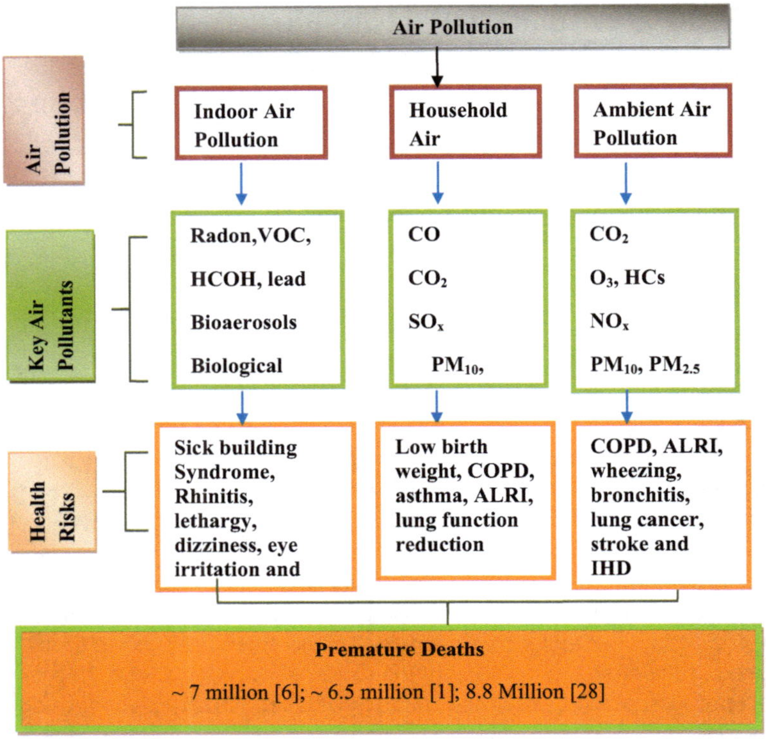

Fig. 1 Type of air pollution, key pollutants, and associated health risks

air-borne particulate pollution [2–5]. Household air pollution is recognized to be the second major environmental health risk factor leading to ill-health effects and majorly responsible for the total burden of disease, accounting to 2.6 million deaths yearly [6]. Worldwide non-communicable diseases are the primary cause of premature mortality, considering this India becomes the first nation to adopt the tenth target dedicated to address the household air pollution in addition to nine specific national targets under the World Health Organization's global action plan for the prevention and control of non-communicable diseases [7]. As per the Lancet commission on pollution and health, acute and chronic exposure to particulate matter has been recognized to be an emerging public health issue amongst lower-middle income countries where it accounts for much of morbidity and over a million deaths annually. A study from North India highlighted that the particulate emissions could be above 10–20 times than the guidelines and standards for outdoor concentrations and worsen indoor pollutant concentrations due to limited and poor ventilation [8, 9]. The vulnerable population—women, elderly peoples, and children—are at most considerable risk to be affected by household air pollution [10].

The main source of household air pollution in developing nations is the use of solid biomass fuels, in inefficient, traditional cooking stoves [11–13]. Incomplete

Fig. 2 Traditional cookstove based on solid biomass fuels being used by rural communities for cooking

burning of solid biomass fuels in kitchens with poor ventilation helps to increase the levels of various harmful household air pollutants. This includes the emissions of particulate matter of variable size (e.g. $PM_{2.5}$, PM_{10}), carbon monoxide (CO), oxides of nitrogen and sulphur (NOx, SOx), black carbon, etc. [14–16]. $PM_{2.5}$ is considered more harmful than PM_{10} due to its deeper penetration into lungs, greater surface area, and more adsorption capacity.

Several studies predicted that emissions due to household energy choice are responsible for 73,000–460,500 premature deaths in India annually [17, 18]. Household energy sector is leading to 67% emissions of $PM_{2.5}$ in India in comparison to other sectors such as land transport, agriculture, and industrial. Recently, a study reported that household energy sector contributes largely to $PM_{2.5}$ emissions, which is the primary cause of various diseases in India [19]. The reduction of about 52% population-weighted annual mean in $PM_{2.5}$ emissions could be attained by simply reducing emissions from the household sector, which attributes 511,000 premature deaths every year. Thus, this could considerably reduce the health burden. Thus, it is essential to target the reduction in $PM_{2.5}$ emissions to address the increasing health implications from increasing air pollution [20, 21]. For effectively mitigating household air pollution, personal exposure studies considering factors such as ventilation and kitchen structure volume that are significant predictors of household air pollution should be studied to further develop exposure assessment models [22]. This would aid in informed interventions, policies, and future epidemiological studies intended for reducing household air pollution exposure.

2 Health Impacts of Household Air Pollution

Majority of rural households being much more dependent on biomass fuels, low-quality inefficient cookstoves, and poorly ventilated kitchens lead to building up of health-damaging pollutants [23–26]. The pollutants from these cookstoves constitute a significant contributor to household air pollution and are the biggest cause of respiratory problems. Also, the use of solid biomass fuels raises the risk of pneumonia by >80% [27]. Other major health issues associated with household air pollution are low birth weight, chronic obstructive pulmonary disease (COPD), acute lower respiratory infections (ALRI), lung cancer, stroke and ischaemic heart disease (IHD), etc. [28]. About 25% of the premature deaths that occur from household air pollution exposure are due to a stroke, which is linked with short-term exposure to particulate matter [29]. Approximately, 1.4 million deaths are caused by stroke and amongst which 50% are in women. 15% of the total deaths are due to ischaemic heart disease and are mostly associated with long-term exposure to particulate matter. Another study highlighted that more than 33% of premature deaths take place from chronic obstructive pulmonary disease in lower-middle income countries, around 17% of lung cancer due to carcinogens released from biomass combustion [30]. However, till date, studies are lacking to establish the association between the effects of variable sizes of particulate matter and the duration of exposure on the occurrence of stroke and related diseases.

The data of premature deaths owing to household air pollution listed by WHO (2012) were used to create a region-specific plot for diseases namely, chronic obstructive pulmonary disease, acute lower respiratory infections, lung cancer, stroke, and ischaemic heart disease as shown in Fig. 3 [31]. Statistics reveal that deaths attributable to household air pollution are significantly elevated in regions of South-East Asia, Western Pacific, low-middle income countries (LMICs). The increasing burden nevertheless indicates that diseases associated with exposure to SBF have become a serious problem throughout India.

WHO has estimated that each year household air pollution is responsible for the deaths of 4.3 million people globally. In India, the incidence of several deaths due to household air pollution was found 18% higher for females than for males (0.26 million:0.22 million) [32]. Long-term impacts of such exposure can hamper the overall development in children and can pose a serious impact on maternal health, especially during pregnancy. Pregnant women are at most risk for infection, malnutrition, anaemia, gestational diabetes, and hypertension, and some of these factors may be exacerbated by household air pollution implicated in their development. The long-lasting health diseases caused due to household air pollution exposure could be

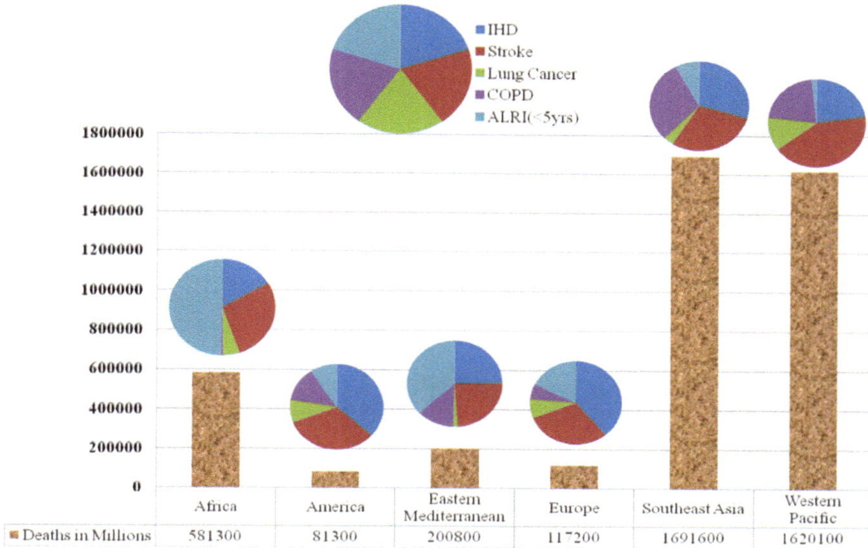

Fig. 3 Total deaths attributable to household air pollution, region-specific

addressed by choosing clean cooking technologies, framing strategic legislative policies and by enhancing effective user-behaviour focused interventions for adoption, and sustained use of clean fuels [33, 34].

Diseases attributable to pollution are reported to cause over 9 million premature deaths, which are thrice the number of deaths caused by AIDS, tuberculosis, and malaria [1]. Another study by the International Energy Agency observed that the deaths related to biomass would increase in the future. These reports also highlight that premature deaths related to biomass uses will continue to increase at a linear rate from 2008 to 2030 as compared to other common diseases like AIDS, malaria, and tuberculosis. Hence, there is a need to understand household air pollution and their exposure better. This can be achieved by generating more regional evidence through air pollution monitoring having coverage of geographic variations. Personal exposure studies can provide better exposure data and associated health risks, but they are costly. This can be achieved by bringing private partners through social corporate responsibilities, which can promote efficient stove and clean fuel usage.

3 Conclusion

Global Burden of Disease and the Lancet Commission on Air Pollution and Health have highlighted that household air pollution is becoming a major health risk. Hence, long-term strategies are required to better address the issue. This can be achieved through proper policies addressing the indoor air quality with the support of the World Health Organization (WHO) guidelines. Reduction in household air pollution will also lead to a decline in ambient air pollution concentration. The success stories of clean household energy and benefits should be communicated to formulate evidence-based policies. Reducing household air pollution through legislation will also reduce the associated health and environmental burden, which will lead to achieving sustainable development goals and accompanying targets finally.

References

1. Landrigan PJ et al (2017) Air pollution and health. Lancet Public Health 2(1):e4–e5
2. Ravindra K (2016) Tackling India's air pollution: the toxic threat that goes beyond the borders of megacities. Policy Forum. http://www.policyforum.net/air-pollutionindia-delhi/
3. Ravindra K, Sidhu MK, Mor S, John S, Pyne S (2015) Air pollution in India: bridging the gap between science and policy. J Hazard, Toxic, Radioact Waste A4015003
4. Ravindra K, Sokhi R, Van Grieken R (2008) Atmospheric polycyclic aromatic hydrocarbons: source attribution, emission factors and regulation. Atmos Environ 42(13):2895–2921
5. Pope CA III, Dockery DW (2006) Health effects of fine particulate air pollution: line that connect. J Air Waste Manag Assoc 56(6):709–742
6. GBD, Risk Factors Collaborators (2017) Global, regional, and national comparative risk assessment of 84 behavioural, environmental and occupational, and metabolic risks or clusters of risks, 1990–2016: a systematic analysis for the Global Burden of Disease Study 2016. Lancet; 390:1345–1422
7. WHO (2017) https://www.who.int/news-room/fact-sheets/detail/household-air-pollution-and-health
8. Ravindra K, Smith KR (2018) Better kitchens and toilets: both needed for better health. Environ Sci Pollut Res 1–4
9. Sidhu MK, Ravindra K, Mor S, John S (2017) Household air pollution from various types of rural kitchens and its exposure assessment. Sci Total Environ 586:419–429
10. Smith K (2013) Biofuels, air pollution, and health: a global review. Springer Science & Business Media
11. Balakrishnan K (2018) Fueling clean household environments. EcoHealth 15(4):876–877
12. Balakrishnan K, Cohen A, Smith KR (2014) Addressing the burden of disease attributable to air pollution in India: the need to integrate across household and ambient air pollution exposures. Environ Health Perspect 122(1):A6
13. Rohra H, Taneja A (2016) Indoor air quality scenario in india—an outline of household fuel combustion. Atmos Environ 129:243–255
14. Ravindra K (2019) Emission of black carbon from rural households kitchens and assessment of lifetime excess cancer risk in villages of North India. Environ Int 122:201–212
15. Ravindra K, Wauters E, Tyagi SK, Mor S, Van Grieken R (2006) Assessment of air quality after the implementation of compressed natural gas (CNG) as fuel in public transport in Delhi. India Environ Monit Assess 115(1):405–417
16. Smith KR, Mehta S, Maeusezahl-Feuz M (2004) Indoor air pollution from household solid fuel use. In: Ezzat M, Lopez AD, Rodgers A, Murray CJL (eds) Comparative quantification of health risks: global and regional burden of disease attributable to selected major risk factors. World Health Organization, pp 1435–1493
17. Lelieveld J, Evans JS, Fnais M, Giannadaki D, Pozzer A (2015) The contribution of outdoor air pollution sources to premature mortality on a global scale. Nature 525:367–371
18. Silva RA, Adelman Z, Fry MM, West JJ (2016) The impact of individual anthropogenic emissions sectors on the global burden of human mortality due to ambient air pollution. Environ Health Perspect 124:1776–1784

19. Conibear L, Butt EW, Knote C, Arnold SR, Spracklen DV (2018) Residential energy use emissions dominate health impacts from exposure to ambient particulate matter in India. Nature Communications 9(1):617
20. Ravindra K, Singh T, Mor S, Singh V, Mandal TK, Bhatti MS et al (2019) Real-time monitoring of air pollutants in seven cities of North India during crop residue burning and their relationship with meteorology and transboundary movement of air. Sci total Environ 690:717–729. Chicago
21. Sharma K, Ravindra K, Mor S, Kaur-Sidhu, M, Sehgal R (2019) Detection and identification of dust mite allergens in the air conditioning filters in Chandigarh, India. Environ Sci Pollut Res 1–10
22. Ravindra K, Rattan P, Mor S, Aggarwal AN (2019) Generalized additive models: building evidence of air pollution, climate change and human health. Environ Int 132:104987
23. Pope D et al (2015) Exposure to household air pollution from wood combustion and association with respiratory symptoms and lung function in nonsmoking women: results from the RESPIRE trial. Guatemala Environ Health Perspect 123(4):285–292
24. Ravindra K, Mittal AK, Van Grieken R (2001) Health risk assessment of urban suspended particulate matter with special reference to polycyclic aromatic hydrocarbons: a review. Rev Environ Health 16(3):169–189
25. Ravindra K, Kaur-Sidhu M, Mor S, John S (2019) Trend in household energy consumption pattern in India: a case study on the influence of socio-cultural factors for the choice of clean fuel use. J Clean Prod 213:1024–1034
26. Ravindra K, Agarwal N, Kaur-Sidhu M, Mor S (2019) Appraisal of thermal comfort in rural household kitchens of Punjab, India, and adaptation strategies for better health. Environ Int 124:431–440
27. Dherani Mukesh, Pope D, Mascarenhas M, Smith KR, Weber M, Bruce N (2008) Indoor air pollution from unprocessed solid fuel use and pneumonia risk in children aged under five years: a systematic review and meta-analysis. Bull World Health Organ 86:390–398C
28. Lelieveld J, Klingmüller K, Pozzer A, Pöschl U, Fnais M, Daiber A, Münzel T (2019) Cardiovascular disease burden from ambient air pollution in Europe reassessed using novel hazard ratio functions. Eur Heart J
29. Sharma D, Ravindra K, Kaur M, Prinja S, Mor S (2020) Cost evaluation of different household fuels and identification of the barriers for the choice of clean cooking fuels in India. Sustain Cities Soc 52:101825
30. Cohen AJ et al (2017) Estimates and 25-year trends of the global burden of disease attributable to ambient air pollution: an analysis of data from the Global Burden of Diseases Study 2015. The Lancet 389(10082):1907–1918
31. WHO (2012) https://www.who.int/sustainable-development/housing/health-risks/household-air-pollution/en/
32. Balakrishnan K, Dey S, Gupta T, Dhaliwal RS, Brauer M, Cohen AJ, Stanaway JD et al (2019) The impact of air pollution on deaths, disease burden, and life expectancy across the states of India: the Global Burden of Disease Study 2017. Lancet Planet Health 3(1): e26–e39
33. Kaur-Sidhu M, Ravindra K, Mor S, John S (2019) Emission factors and global warming potential of various solid biomass fuel-cook stove combinations. Atmos Pollut Res
34. Kaur-Sidhu M, Ravindra K, Mor S, John S, Aggarwal AN (2019) Respiratory health status of rural women exposed to liquefied petroleum gas and solid biomass fuel emissions. Air Soil Water Res 12:1178622119874314

Characteristics of PM from Different South Indian Cooking Methods and Implications in Health Effects

Yaparla Deepthi, S. M. Shiva Nagendra and Sathyanarayana N. Gummadi

Abstract Indoor air pollution (IAP) predominantly contributed from biomass burning in rural households is a major health hazard. Cooking activities are significant sources of indoor particulate matter (PM). The present study focuses on characterising PM emissions from different cooking methods that are primarily prepared in rural areas of South India, in a simulated kitchen relying on biomass as fuel and estimation of respiratory dosage. Controlled experiments were carried out to study PM concentrations generated while performing different cooking methods including boiling (rice, *urad dal* and preparation of tea) and pan-frying (wheat *roti* and *omlette*). Multiple Particle Path Dosimetry (MPPD) was used to estimate deposition fractions in head, tracheobronchial and pulmonary regions of the human respiratory tract (HRT) for women. Further, PM dosage was assessed by entering the captured PM measurements and evaluated amongst different cooking methods. PM concentrations from pan-frying were ~1.6 times greater than boiling, primarily due to usage of oil for frying. Furthermore, pan-frying displayed higher dosage (412–2240 μg) compared to the boiling (258–1119 μg). However, *urad dal* displayed extreme amplification of 8.7 times than preparation of tea due to longer cooking duration. It is evident from above results that cooking methods are major attributes impacting IAP in rural areas with severe health impacts.

Keywords Indoor air pollution · Biomass · Cooking methods · MPPD · Dosage

1 Introduction

One of the foremost health hazards in India is Indoor air pollution (IAP). As per World Health Organisation (WHO), India carries the largest burden of diseases due to IAP [1] which is designated as one of the four most critical global environmental

Y. Deepthi · S. M. Shiva Nagendra (✉)
Department of Civil Engineering, Indian Institute of Technology Madras, Chennai 600 036, India
e-mail: snagendra@iitm.ac.in

Y. Deepthi · S. N. Gummadi
Department of Biotechnology, Indian Institute of Technology Madras, Chennai 600 036, India

A. Sharma et al. (eds.), *Indoor Environmental Quality*,
Lecture Notes in Civil Engineering 60,
https://doi.org/10.1007/978-981-15-1334-3_5

problems along with ischaemic heart disease (IHD) and stroke and chronic obstructive pulmonary disease (COPD). The most common source of IAP in rural India is burning of solid fuels including biomass and coal for cooking activities [2]. It is estimated that of the 2.8 billion biomass users across the world, India alone is home to more than 0.82 billion people, a majority of which comes from its rural areas [3]. Fuel wood (FW) is predominantly used as domestic fuel in rural India. Additionally, assessing IAP activity wise, particulates in rural households are mostly contributed from biomass combustion for cooking, smoking and re-suspension of household dust and through in-filtration of outdoor sources like agricultural residue burning and unpaved roads. Fine and ultrafine particulate matter (PM) generated in these activities have greater health consequences due to higher mass concentration levels and the presence of toxic substances in its composition [4]. Amongst the various activities, cooking is one of the major sources contributing towards fine and ultrafine PM. Past studies reported high levels of PM mass and number concentrations, chemical and morphological compositions from cooking activities across the globe [5, 6]. Furthermore, these PM characteristics depend on various attributes like cooking method, cooking style, type of the energy source and its quality, varied oil types, raw materials, additives, cooking pan, position of cooking pan and burner size [5, 7, 8]. However, most of these studies focused on cooking activity performed on gas cooking either in controlled cooking experiments or in real-world residential kitchens [5, 7, 9].

In addition, cooking methods differ across the world which in turn impacts the PM characteristics. Wet cooking, for example, includes cooking that primarily requires water, i.e. boiling, stewing and steaming and frying consists of stir-frying, pan-frying and deep-frying, while dry cooking comprises broiling, grilling, oven baking, toasting and microwaving that are the most common cooking methods [10]. PM_{10} and $PM_{2.5}$ concentrations during steaming were low followed by boiling and highest from frying in a study performed by Lee et al. [11]. See and Balasubramanian [5] studied particle number concentrations for boiling, steaming, stir-frying, pan-frying and deep-frying and observed that frying resulted in higher concentration compared to steaming and boiling. Also, Alves et al. [12] measured $PM_{2.5}$ concentration for different cooking methods and a similar trend was observed, i.e. frying > grilling > stewing > boiling. In a review study by Torkmahalleh et al. [10], various studies were analysed and concluded that low particle concentrations were released during boiling and steaming compared to oil-based frying. However, very few studies compared these cooking methods for Indian cooking styles [5, 13]. Thus, it is very essential to evaluate PM characteristics with different cooking methods so as to assess the health risks associated with the cooking activities.

PM measurements in conjunction with the analysis of deposition of particulates those inhaled in vulnerable regions of respiratory systems comprehend particulate pollution and its allied health hazards [14]. Various models have been reported to estimate particulates deposition fraction in HRT to predict deposition plus clearance [15–17]. However, the Multiple Particle Path Dosimetry (MPPD) model (developed by the Environment and Chemical Industry Institute of Toxicology (CIIT, USA) and

Dutch National Institute for Public Health) is widely used as it is more representative for studies associated with particle dosimetry as it contemplates asymmetric branching pattern in respiratory systems [18–20].

Several past studies have been carried out to comprehend PM characteristics from different cooking activities across the globe [6–9, 12] but a few numbers of studies for Indian cooking styles [5]. Since the cooking characteristics significantly depend on cooking method, cooking style, varied oil types, raw materials and additives, the disparity in terms of significance of the studies is the subject for the present study. Furthermore, only handful studies are conducted on woman specifically, during cooking, for particulate respiratory dosage [20]. Hence, the objectives of the present study are to evaluate PM characteristics for different cooking methods including boiling (rice, *urad dal* and preparation of tea) and pan-frying (wheat *roti* and *omlette*) with constant cooking conditions and fuel wood, followed by assessment of respiratory dosage for women and associated health impacts.

2 Methodology

2.1 Monitoring Campaign, Site and Instruments

The samplings were carried out in a controlled room (3 m × 3 m × 3 m) with two windows (1.15 m × 1.15 m) and one door (2.3 m × 1.10 m) (Fig. 1) located in IIT Madras campus. Fuel wood was used for cooking which is predominantly used in rural households of South India. No wood was present and all the glass windows were kept closed during the experiments so as to control ventilation. Thus, the PM concentrations monitored are primarily from cooking activities.

PM mass and number distribution ranging from 300 nm to 20 μm were measured using 16-channel GRIMM optical particle counter (1.108) (Fig. 2a). The instrument uses a semiconductor laser as the light source as it works on the principle of light scattering technology (Fig. 2b). Periodic calibration of instrument was performed to ensure precision of data collected with reproducibility of ±2%. Before each monitoring, all the necessary flow and background checks of the instrument were performed that was kept at 1.5 m away from cooking activity at respirable height (1 m) during each cooking experiment. Further, using instrument software, the PM data was extracted at 1-min resolution and subsequently data was analysed using Microsoft Excel (MS Office 2013) and origin pro (8.5 version).

2.2 Sample Collection

The experiments carried out in this study were designed to simulate the PM concentrations (PM_{10}, $PM_{2.5}$ and PM_1) arising in domestic Indian cooking style primarily in rural households of South India. The cooking methods considered for the study

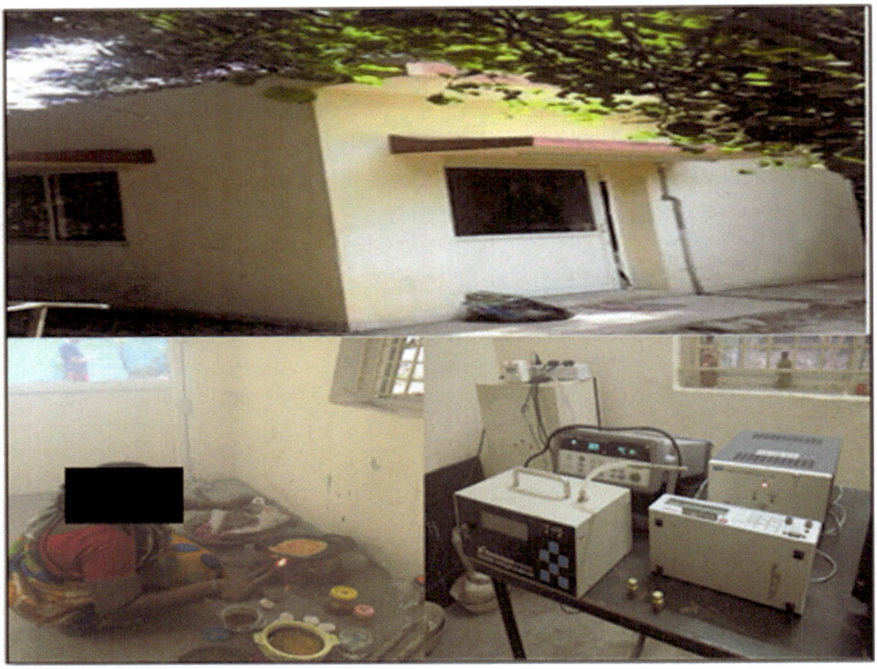

Fig. 1 Photograph of controlled room along with women performing cooking activity

include boiling, i.e. rice, preparation of tea and *dal* and pan-frying, i.e. wheat *roti* and *omlette* (Table 1). Above cooking methods were duplicated keeping fuel wood and quantity prepared as constants. Each experiment was performed on a single day so as to disperse the air that would have attributed from the previous experiments.

3 Particle Dosimetry Model

Deposition fraction (DF) is the most common metric used to determine depositions in head (H), tracheobronchial (TB) and pulmonary (P) regions of HRT. DF is a function of mass median aerodynamic diameter (MMAD), tidal volume, breathing rate, geometric standard deviation (GSD) and anatomy of lungs. DF could be calculated by different mathematical models like International Commission on Radiological Protection (ICRP) model, the National Council on Radiation Protection and Measurements (NCRP) and MPPD. ICRP 66 is a semi-empirical approach based on algebraic equations resulting from experimental and theoretical results developed in 1994. It is a single path model with simplified morphometry which is a major limitation [17, 21]. MPPD model (MPPD v 3.04, https://www.ara.com/products/multiple-path-particle-dosimetry-model-mppd-v-304) in contrast is single as well as

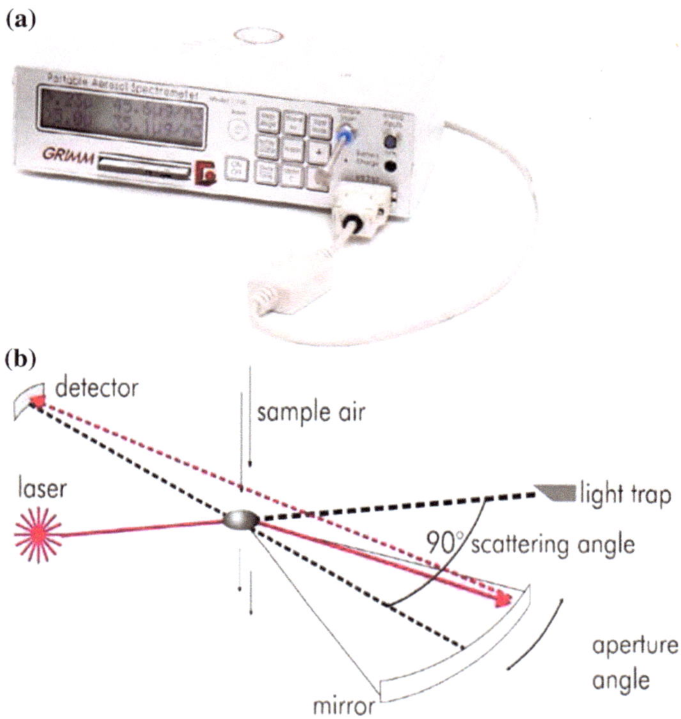

Fig. 2 Instruments used for monitoring **a** GRIMM (Model 1.108) and **b** GRIMM working principle

Table 1 Experiment procedure for different cooking methods

S. No.	Cooking method	Item	Cooking time (min)
1	Boiling	Rice	22
2	Boiling	Preparation of tea	10
3	Boiling	Dal	42
4	Frying	Wheat roti	23
5	Frying	Omlette	20

multi-path model. This model relies on real measurements of single airways and corresponding asymmetric branching structure of the lung [22]. Consequently, it leads to a representative estimation of DFs in all regions of the lung and provides the prospect to have detail assessment of dosage and subsequently associated health risks. Thus, for the present study, the MPPD model was considered. So as to assess regional and whole lung depositions, the combination of various deposition mechanisms like inertial impaction, diffusion and sedimentation was considered [23–25]. Stochastic lung model (60th percentile) was considered as it is the most accurate models for

Table 2 Physiological/morphological parameters for adult during lightweight activity

Activity	Time (min)	Height (cm)	Body weight (kg)	BF (breaths/min)	TV (ml)	URT (ml)
Light indoor activity (Cooking)	10–42	151	44.2	22	820	2110

human depositions. The input parameters included (a) particle properties (density, size and shape distribution) (b) specific activity (cooking: light indoor activity) (c) women characteristics such as age, gender, weight and height (d) respiratory physiological parameters like tidal volume (TV), breathing frequency (BF), functional residual capacity (FRC) and upper respiratory tract (URT) volume. For each cooking method, MMAD and GSD were calculated [20]. Additionally, only nasal breathing scenario with uniformly expanding flow and 0.5 inspiratory fraction was assumed. Table 2 provides the input parameters obtained from the literature specific to women [26–28].

Dosage is the amount of pollutant a person inhales during exposure to different activities (cooking) for certain duration of time. It largely depends on deposition fraction (dependent on particle size), the pollutant concentration (particulate matter), duration of exposure and breathing pattern. The DF obtained from MPPD model during each cooking method is used to estimate equivalent dose (expressed in µg) by Eq. (1)

$$\text{Dose} = \text{PM} \times \text{DF} \times \text{TV} \times f \times t \tag{1}$$

'DF' is deposition fractions (dimensionless), 'PM' is cooking concentration (µg/m^3), 'TV' is tidal volume of women during cooking activity (m^3/breath), 'f' is breathing frequency of women during cooking activity (breaths/min), and 't' is cooking time exposure (min) [20].

4 Results and Discussion

4.1 PM Mass Concentrations for Different Cooking Methods

The PM_{10}, $PM_{2.5}$ and PM_1 mean mass concentrations released during different cooking methods are presented in Fig. 3. PM concentrations from pan-frying were ~1.6 times greater than boiling, primarily due to usage of oil for frying [29]. The following trend was observed, wheat *roti* > *dal* > *omlette* > rice > tea (Table 3).

These values were comparable to the past studies carried out across the globe. The trend was similar to the studies carried out by Lee et al. [11], See and

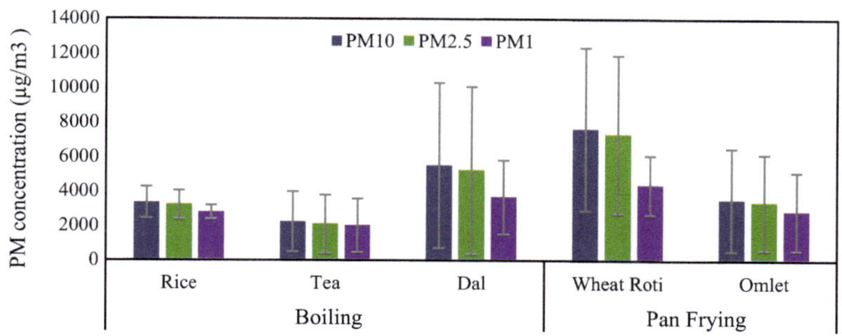

Fig. 3 PM mass concentrations for different cooking methods

Table 3 Trends of PM_{10}, $PM_{2.5}$ and PM_1 during various items of cooking

Cooking item	PM_{10} ($\mu g/m^3$)	$PM_{2.5}$ ($\mu g/m^3$)	PM_1 ($\mu g/m^3$)
Rice	3350	3211	2776
Preparation of tea	2236	2087	2004
Dal	5495	5215	3669
Wheat *roti*	7615	7295	4374
Omlette	3524	3365	2840

Balasubramanian, [5], Alves et al. [12], Abdullahi [13], i.e. frying followed by boiling.

4.2 *PM Number Concentrations for Different Cooking Methods*

Quantification and sub-classification of coarse and fine particles in the size ranges considered by Deepthi et al. [20] in their study (Table 4) were used to understand the

Table 4 Particle size-based concentrations for different cooking methods

Type of house	0.3–0.9 μm (particles/cm^3)	0.9–1.8 μm (particles/cm^3)	1.8–4.5 μm (particles/cm^3)
Rice	16.7×10^6	3.4×10^4	2.5×10^3
Tea	12.6×10^6	5.8×10^3	1.7×10^3
Dal	16.2×10^6	2.4×10^5	7.4×10^3
Wheat *roti*	16.1×10^6	6.1×10^5	1.3×10^4
Omlette	16.1×10^6	4.7×10^4	2.9×10^3

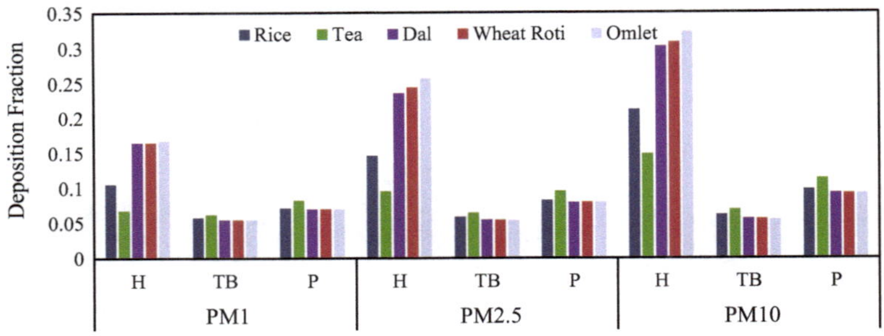

Fig. 4 PM deposition fractions in HRT for different cooking methods

particle concentrations for different cooking methods. It was observed that number concentrations were comparable in size range 0.3–0.9 μm for all the cooking methods. However, concentrations from wheat roti were one power higher compared to other cooking methods for 0.9–1.8 and 1.8–4.5 μm. As, *roti* preparation itself results in increased generation of the number of particles emitted apart from oil frying. These number concentrations were comparable to past studies reported in Chinese cooking methods by Zhang et al. [8] and See and Balasubramanian [5]. Apart from this, number concentrations also establish the size-dependent deposition in the different regions of HRT.

5 PM Deposition Fraction in HRT

For all the particles sizes and all the cooking methods, it was observed that the DFs were considerably higher in H region of HRT (Fig. 4). The variance in DFs with variable particle sizes was primarily reliant on the deposition mechanisms in the HRT, i.e. inertial deposition, gravitational setting and diffusional deposition. The difference in DFs is also majorly reliant on breathing pattern and ventilation that **vary** with age (women) and activities (cooking).

6 PM Dosage in HRT

Dosages were calculated for different cooking methods for PM_1, $PM_{2.5}$ and PM_{10} that are summarised in Fig. 5. Pan-frying demonstrated higher dosage (412–2240 μg) compared to the boiling (258–1119 μg). Amongst each method, dal unveiled extreme amplification of 8.7 times compared to preparation of tea due to longer cooking duration. The trend observed was *dal* > wheat *roti* > *omelette* > rice > tea.

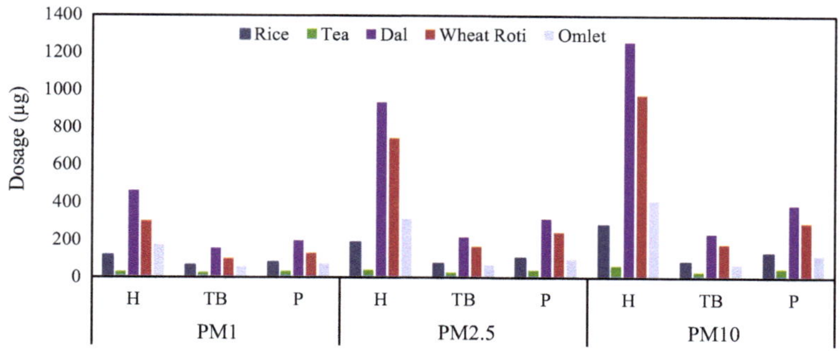

Fig. 5 PM_1, $PM_{2.5}$ and PM_{10} dosage in women

7 Conclusions

The present study records PM_1, $PM_{2.5}$ and PM_{10} levels reported from various cooking methods that are typically prepared in rural households of South India and also estimation of respiratory dosage for women in different regions, i.e. H, TB and P of HRT. It was observed that PM concentrations were highest during pan-frying and lowest from boiling. Additionally, cooking activities emitted millions of aerosol particles (~10^6 particles/cm^3) from all the cooking methods. Furthermore, the study provides a comprehension to amount of particulates deposition for women in different regions of HRT from varied cooking methods. The results revealed that pan-frying displayed higher dosage (412–2240 µg) compared to boiling (258–1119 µg). Thus, it was comprehended once again that different cooking methods are major attributes impacting IAP and have health implications which need serious attention.

References

1. World Health Organization (2009) Global health risks: mortality and burden of disease attributable to selected major risks. World Health Organization. vi, 62 p
2. Smith KR, Aggarwal AL, Dave RM (1983) Air pollution and rural biomass fuels in developing countries: a pilot village study in India and implications for research and policy. Atmos Environ (1967) 17(11):2343–2362
3. Rohra H, Taneja A (2016) Indoor air quality scenario in India—an outline of household fuel combustion. Atmos Environ 129:243–255
4. Andreae MO, Merlet P (2001) Emission of trace gases and aerosols from biomass burning. Glob Biogeochem Cycles 15(4):955–966
5. See SW, Balasubramanian R (2006) Physical characteristics of ultrafine particles emitted from different gas cooking methods. Aerosol Air Qual Res 6(1):82–92
6. Li YC, Shu M, Ho SSH, Wang C, Cao JJ, Wang GH ,... Zhao XQ (2015). Characteristics of $PM_{2.5}$ emitted from different cooking activities in China. Atmos Res 166:83–91
7. See SW, Balasubramanian R (2008) Chemical characteristics of fine particles emitted from different gas cooking methods. Atmos Environ 42(39):8852–8862

8. Zhang N, Han B, He F, Xu J, Zhao R, Zhang Y, Bai Z (2017) Chemical characteristic of $PM_{2.5}$ emission and inhalational carcinogenic risk of domestic Chinese cooking. Environ Pollut 227:24–30
9. Zhang Q, Gangupomu RH, Ramirez D, Zhu Y (2010) Measurement of ultrafine particles and other air pollutants emitted by cooking activities. Int J Environ Res Public Health 7(4):1744–1759
10. Torkmahalleh MA, Gorjinezhad S, Unluevcek HS, Hopke PK (2017) Review of factors impacting emission/concentration of cooking generated particulate matter. Sci Total Environ 586:1046–1056
11. Lee S, Li W, Chan L (2001) Indoor air quality at restaurants with different styles of cooking in metropolitan Hong Kong. Sci Total Environ 279:181–193
12. Alves C, Duarte M, Nunes T, Moreira R, Rocha S (2014) Carbonaceous particles emitted from cooking activities in Portugal. Global NEST J. 16:411–419
13. Abdullahi LK (2017) Characterisation of particulate matter emissions from cooking (Doctoral dissertation, University of Birmingham)
14. Varghese SK, Gangamma S, Patil RS, Sethi V (2005) Particulate respiratory dose to Indian women from domestic cooking. Aerosol Sci Technol 39(12):1201–1207
15. Yeh HC, Schum GM (1980) Models of human lung airways and their application to inhaled particle deposition. Bull Math Biol 42(3):461–480
16. Koblinger L, Hofmann W (1990) Monte Carlo modeling of aerosol deposition in human lungs. Part I: Simulation of particle transport in a stochastic lung structure. J Aerosol Sci 21(5):661–674
17. ICRP (1994) Human respiratory tract model for radiological protection, ICRP Publication 66. Ann ICRP. 24:1–3
18. Hussain M, Madl P, Khan A (2011) Lung deposition predictions of airborne particles and the emergence of contemporary diseases, Part-I. Health 2(2):51–59
19. Menon JS, SM SN (2017) Statistical distribution and particle dosimetry models to estimate personal exposure at urban sidewalks of tropical climate. Sustain Cities Soc 40:254–265
20. Deepthi Y, Nagendra SS, Gummadi SN (2019) Characteristics of indoor air pollution and estimation of respiratory dosage under varied fuel-type and kitchen-type in the rural areas of Telangana state in India. Sci Total Environ 650:616–625
21. Jarvis NS, Birchall A, James AC, Bailey MR, Dorrian M-D (1996) LUDEP 2.0. Personal computer program for calculating internal doses using the ICRP Publication 66 respiratory tract model. Chilton, NRPB-SR287
22. Asgharian B, Hofmann W, Bergmann R (2001) Particle deposition in a multiple-path model of the human lung. Aerosol Sci Technol 34(4):332–339
23. Manigrasso M, Buonanno G, Fuoco FC, Stabile L, Avino P (2015) Aerosol deposition doses in the human respiratory tree of electronic cigarette smokers. Environ Pollut 196:257–267
24. Manigrasso M, Buonanno G, Stabile L, Morawska L, Avino P (2015) Particle doses in the pulmonary lobes of electronic and conventional cigarette users. Environ Pollut 202:24–31
25. Manigrasso M, Vernale C, Avino P (2017) Traffic aerosol lobar doses deposited in the human respiratory system. Environ Sci Pollut Res 24(16):13866–13873
26. Vijayan VK, Kuppurao KV, Venkatesan P, Sankaran K, Prabhakar R (1990) Pulmonary function in healthy young adult Indians in Madras. Thorax 45(8):611–615
27. Jain SK, Ramiah TJ (1967) Influence of age, height and body surface area on lung functions in healthy women 15–40 years old. Indian J Chest Dis 9(1):13
28. IAEA (1998) Compilation of anatomical, physiological and metabolic characteristics for a reference Asian man. Vol. 1 and 2, Report IAEA-TECDOC-1005, International Atomic Energy Agency, Vienna, Austria
29. Buonanno G, Johnson G, Morawska L, Stabile L (2011) Volatility characterization of cooking-generated aerosol particles. Aerosol Sci Technol 45(9):1069–1077

Chamber Studies for Indoor Air Quality Modeling and Monitoring

Sumanth Chinthala, Sunil Gulia and Mukesh Khare

Abstract Evaluating the status of indoor air quality using scientific techniques has become a necessity at both urban and rural habitats. The process generally involves monitoring of pollutants, investigation of its dispersion characteristics, formation and destruction of pollutants, and rate of addition and removal from the sources and sinks, respectively. In order to assess the above, the usage of sophisticated instruments or low-cost sensors has become a prerequisite. However, their unavailability and affordability have a significant effect on the indoor air quality studies at various scales. To an extent, these studies can be performed using the computational fluid dynamics (CFD) models which can simulate the pollutant dispersion characteristics based on predefined numerical solvers. Moreover, information about the fate and transport of the pollutant in the real time at the full-scale level remains unexplained. Chamber studies enable us to supplement the monitoring studies conducted at full-scale levels along with the monitoring studies and CFD simulations. The current paper discusses the various types of indoor air quality chambers and their applications in the investigation of different air quality parameters. The paper also presents the case studies in development of ECO-SEE wall panels and the ability of the construction materials to absorb pollutants.

Keywords Indoor air quality · CFD · Chambers · Modeling · Monitoring

S. Chinthala (✉)
National Institute of Technology Warangal, Warangal, India
e-mail: sumanthchinthala@nitw.ac.in

S. Gulia
National Environmental Engineering Research Institute, Delhi Zonal Centre, New Delhi, India

M. Khare
Civil Engineering Department, Indian Institute of Technology Delhi, New Delhi, India

A. Sharma et al. (eds.), *Indoor Environmental Quality*,
Lecture Notes in Civil Engineering 60,
https://doi.org/10.1007/978-981-15-1334-3_6

1 Introduction

The indoor air quality in buildings is dependent on both the nature of air movement within the building systems and the nature and location of contaminant sources. Emission of indoor air pollutants like volatile organic compounds has direct influence on people's well-being and health. High exposure of indoor air pollutants concentration may cause conjunctival irritation, nose and throat discomfort headache, allergic skin reaction, dyspnea declines in serum cholinesterase levels, nausea, emesis, epistaxis, fatigue, and dizziness [1]. Further, these higher concentrations cause various health problems that may result into "sick building syndrome." The indoor air pollutants can originate from outdoor sources (e.g., traffic and other forms of combustion (CO, NO_2, SO_2) or from indoor sources, such as occupants and their activities, tobacco smoke, electronic equipment, cleaning products or heating, ventilating, air-conditioning (HVAC) systems, and building and furnishing materials. Additionally, volatile organic compounds (VOCs) in indoor air may be found in paints, wood preservatives, aerosol sprays, cleansers and disinfectants, moth repellents and air fresheners, stored fuels and automotive products, and building materials [1]. Moreover, the degradation and deterioration of artificial and natural) building materials may also release harmful pollutants. Hence, it is important to know the influence of these pollutants not only on the health of the humans but also on the building materials. In this context, the study of the transport, deposition, and re-emission of these materials in environmental chambers allows us to predict their behavior in confined indoor spaces. Nowadays, many advanced materials are used specially in construction and restoration works in indoor spaces. Environmental chambers have to be designed to perform controlled experiments to evaluate the environmental variables which can be used in the development of indoor air quality models. The developed models shall be validated by performing full-scale testing under controlled environments to give a complete overview of the pollutant transfer within the domain.

2 Environmental Chambers: An Overview

An environmental chamber is an enclosure used to test the effects of specified environmental conditions on species, industrial products, materials, and electronic devices and components. They artificially replicate the conditions under which the specimen and their components might be exposed. They are also employed to accelerate the effects of exposure to the environment to predict future scenarios. These scenarios may include large, swift variations in temperature, very high or low relative humidity, high and low pressures, etc. Few studies conducted on monitoring pollutants using environmental chambers are listed in Table 1.

Table 1 Chamber studies conducted for monitoring air pollutants

S. No.	Chamber description	Pollutants monitored	References
1	Stone exposure rig	SO_2	[2]
2	Atmospheric flow chambers	SO_2, NO_2, NO, HCL	[3]
3	NO_2 exposure apparatus	NO_2	[4]
4	Environmental recirculation flow chamber	SO_2, NO_2, O_3	[5]
5	Atmospheric flow chamber	SO_2	[6]
6	Small-scale chamber	VOCs	[7]
7	Acrylic chamber	CO_2	[8]

3 Chamber Studies for VOCs Emitted from Building Materials

Building materials may also affect the transport and removal process of indoor VOCs by their *adsorption* and *desorption* properties. Re-emission of adsorbed VOCs may significantly increase indoor VOC concentrations for months or years after a source event [9]. The sorption properties on a material for VOC can be evaluated in two ways, i.e., experimental investigation and emission modeling [10, 11]. In principle, the experimental measurement methods provide more realistic results than the mathematical modeling. However, such methodologies require expensive well-controlled instrumentation and time. Many researchers have addressed the importance of simulating VOC emissions from building materials and furnishings using mathematical models.

4 Interaction of VOCs with Building Materials

The interaction process between materials and pollutant involves transfer of pollutants (from the bulk air phase to the material–air interface), dynamic exchange of VOCs between air phase and material surface, and diffusion of VOCs in the interior of the material, if it is permeable [12]. Further, the exchange of VOCs at the air–material interphase may involve physisorption and chemisorption.

Moreover, studies conducted by Tichenor et al. [13] on impacts of VOCs on materials indicate that VOC concentration has no significant influence on the sorption capacity of a material-VOC combination. Hence, sorption tests may be preferably conducted under relatively higher concentrations to improve accuracy. The studies also found that the adsorption–desorption and rate of tetrachloroethylene on carpet at 35 °C was significantly higher than those at 23 °C.

Generally, the sorption models are classified as equilibrium-interface models and first-order rate models. For the interfacial mass transfer, a constant coefficient is used to represent the ratio between the interfacial VOC concentrations for the two phases.

On the other hand, the first-order rate model assumes that adsorption and desorption processes are, respectively, proportional with the VOC concentration in the air and on the surface of the material. At a later stage, Zhang et al. [7] have reviewed various sorption models and their explaining their assumptions and applications. Table 2 describes the detailed assessment of various sorption models.

5 Monitoring and Modeling of VOCs Using Chamber Studies

The methods of studying characteristics of VOC sources and sinks mainly fall into two categories: experimental investigation and emission modeling [10, 11]. In principle, the experimental measurements provide the most realistic results than the modeling results. However, it requires expensive and well-controlled instrumentation. The experiments are set up using a small-scale or a full-scale stainless steel or glass chamber to measure the sorption properties of a test material. The tests are usually conducted under a set of specific environmental conditions (e.g., 25 °C, 50% RH, and 1 ACH). The experiments are generally completed in two phases, the dynamic adsorption phase and the dynamic desorption phase. In adsorption phase, the compounds generated from the pollutant generator are carried by the conditioned, clean air from the conditioner to the chamber containing the test specimen. The pollutants concentration in the chamber is measured by analyzing air samples taken from the chamber exhaust. After the system reaches an apparent equilibrium (the concentrations at the chamber exhaust does not increase any more), the dynamic desorption phase starts whereby the pollutant supply is stopped and the chamber is continuously flushed out by the clean air.

Owing to the limitation of the experimental approach, many researchers have addressed the importance of simulating VOC emissions from building materials and furnishings using mathematical models. In literature, the sorption model is generally two types, i.e., first-order adsorption/desorption rate models and equilibrium-interface models [11]. The linear Langmuir model is probably the most widely used sorption model. It is based on physisorption process and considers only the relatively fast surface sorption process. It does not consider the slow diffusion of pollutant inside the material. The linear Langmuir model is described by the following equation [13].

$$\frac{\mathrm{d}M}{\mathrm{d}t} = k_{\mathrm{a}}C - k_{\mathrm{d}}M \tag{1}$$

where

$\mathrm{d}M/\mathrm{d}t$ = net mass rate of change of VOCs adsorbed on the material surface (μg m^{-2} h^{-1}); C = concentration in chamber (μg/m^3); k_{a} = adsorption rate coefficient (m h^{-1}); k_{d} = desorption rate coefficient (h^{-1}); and M = mass of pollutant per unit area on material surface (μg m^{-2}).

Table 2 Detailed assessment of various sorption models

Models	Governing equations	Parameters	a	b	c	d	e	References
Linear Langmuir	$\mathrm{d}M/\mathrm{d}t = k_\mathrm{a}C_a - k_\mathrm{d}M$	$k_\mathrm{a}, k_\mathrm{d}$	N	Y	N	C-F	Y	[13]
K-diffusion model	$\mathrm{d}W_1/\mathrm{d}t = k_3VC_a - k_4W_1$ $\mathrm{d}W_2/\mathrm{d}t = k_5VC_a - k_6W_2$	k_3, k_4, k_5, k_6	N	Y	N	C-F	Y	[14–16]
Sorption–diffusion hybrid model	$\partial C_m(x,t)/\partial t = D_m\dfrac{\partial 2C_m(x,t)}{\partial x2}$ $C_m(+0, t) = C_a(t)$	D_m	N	Y	Y	C-F	Y	[17]
Sorption–diffusion hybrid model	$\partial C_m(x,t)/\partial t = D_m\dfrac{\partial 2C_m(x,t)}{\partial x2}$ $\mathrm{d}W(t)/\mathrm{d}t = k_3VC_a(t) + AD_m$ $\partial Cm(x,t)/\partial x\vert_{x=+0} - k_4W(t)$ $C_m(+0, t) = W(t)/A$	D_m, k_3, k_4	N	Y	Y	C-F	Y	[17]

[a]Consideration of VOC transport in the air
[b]Consideration of surface sorption
[c]Consideration of VOC diffusion in the material
[d]Determination of model parameters: C-F: curve fitting
[e]Application for IAQ prediction and coupling with CFD code

Table 3 Application of CFD for adsorption and environmental chamber studies

S. No.	Application	References
1	Simulation of binary adsorption	[12]
2	Adsorption/desorption processes in carbon	[19]
3	Dispersion of exhaled air in a ventilation room	[20]
4	CFD simulation of respiration	[21]
5	CFD in a test chamber	[18]
6	Modeling adsorption using CFD for cooling system	[22]
7	CFD simulation for adsorption separation in chromatography	[23]
8	Modeling protein adsorption using CFD	[24]

It is difficult to measure the values of k_{a} and k_{d} experimentally. However, researchers used statistical tools to estimate the value of these coefficients based on monitoring data of concentration and emission rate.

6 CFD Simulations—An Overview

Computational fluid dynamics (CFD) has been widely used as a method of simulating room airflow, studying indoor environment issues, and producing data that may be otherwise difficult to obtain through in situ measurements [18]. The application of CFD for simulating the adsorption and desorption kinetics has been investigated in various studies [12, 19]. Additionally, the dispersion of the pollutants in the controlled environmental chambers has also been investigated widely [20, 21]. Further, White, 2012, has conducted a study comparing CFD simulations with experimental results (Table 3).

7 Adsorption–Desorption Experiments in 2 L Chambers and Simulation Studies in ECO-SEE

In order to determine the behavior of test materials when exposed to VOCs, "2 L" cylindrical chambers developed by Building Research Establishment (BRE), London, were used (actual internal volume 1.7–1.9 L). Due to the different types of test materials, use of such chambers was preferred to the use of other methods such as Field and Laboratory Test Cell (FLEC).

The VOCs selected for the studies were: toluene (a surrogate for benzene), limonene (a ubiquitous terpene used in consumer products as a fragrance), dodecane (a hydrocarbon representing those present in fuels and as a solvent in decorative finish products), and formaldehyde (a VVOC).

The same VOCs and their adsorption and desorption on the test materials were performed using FLUENT 14.0. The boundary conditions and the duration of the simulations were same as that of the experimental conditions. The simulations are primarily targeted toward numerically predicting behavior of materials for selected pollutants, i.e., whether they act as sink or re-emitter. The behavior of the materials under various conditions has been described in [25, 26]. The obtained results indicate that the combined results obtained from chamber studies, CFD simulations, and the numerical studies are essential to estimate the behavior of pollutants and materials in the indoor environments.

Acknowledgements The results presented in this paper are part of the ECO-SEE project which has received funding from the European Union's Seventh Framework Programme for research, technological development and demonstration under grant agreement No. 609234 (www.eco-see.com).

References

1. USEPA (2017) Volatile organic compounds' impact on indoor air quality. https://www.epa.gov. Accessed on 10 Mar 2019
2. Brien PFO, Bellb E, Orra TLL, Cooperc TP (1995) Stone loss rates at sites around Europe. Sci Total Environ 167(1–3):111–121. https://doi.org/10.1016/0048-9697(95)04574-K
3. Johnson JB, Haneef SJ, Hepburn BJ, Hutchinson AJ, Centre P (1990) Laboratory exposure systems to simulate atmospheric degradation of building stone. Atmos Environ Part A. General Topics 24(10):2585–2592. https://doi.org/10.1016/0960-1686(90)90136-B
4. Finch BSC (2011) Modeling transport and protein adsorption in microfluidic systems. Doctoral Thesis. College of Sciences, University of Central Florida, Orlando, Florida, United States, p 133
5. Chinthala S, Gulia S, Silva D, Rana C, Mower K, Dengel A, Khare M (2016) Emission characteristics of eco-friendly sustainable building construction material-Part II. In: Presented at Indoor air conference, Ghent, Belgium. July 3–8 (2016)
6. Gulia S, Chinthala S, Silva D, Rana C, Mower K, Dengel A, Khare M (2016) Emission characteristics of eco-friendly sustainable building construction material-Part I. In: Presented at Indoor air conference, Ghent, Belgium, July 3–8 (2016)
7. Zhang JS, Zhang JS, Chen Q, Yang X (2002) A critical review on studies of volatile organic compound (VOC) sorption on building materials. ASHRAE Trans 108(1):162–174
8. Gupta A, Nema AK (2011) CO_2 sequestration by alkaline wastes for potential application in the landfill environment. In: International conference on environmental and computer science IPCBEE vol 19. IACSIT Press, Singapore
9. Nielsen PA (1987) Potential pollutants—their importance to the sick building syndrome—and their release mechanism. Proc Indoor Air 87(2):598–602
10. Zhu JP, Zhang JS, Shaw CY (2001) Comparison of models for describing measured VOC emissions from wood based panels under dynamic chamber test condition. Chemosphere 44:1253–1257. https://doi.org/10.1016/s0045-6535(00)00338-6
11. Zhao DY, Little JC, Hodgson AT (2002) Modeling the reversible diffusive sink effect in response to transient contaminant sources. Indoor Air 12:184–190. https://doi.org/10.1034/j.1600-0668.2002.01116.x
12. Yang X, Chen Q, Zhang JS (2001) A mass transfer model for simulating on VOC sorption on building material. Atmos Environ 35:1291–1299. https://doi.org/10.1016/S1352-2310(00)00397-6

13. Tichenor BA, Guo Z, Dunn JE, Sparks LE, Mason MA (1991) The interaction of vapor phase organic compounds with indoor sinks. Proc Indoor Air 1:23–35
14. Colombo A, De Bortoli M, Knoppel H et al (1993) Vapor deposition of selected VOCs on indoor surface materials in test chambers. In: Proceedings of Indoor Air'93, vol 2, pp 407–412
15. Colombo A, De Bortoli M, Knoppel H et al (1993) Adsorption of selected volatile organic compounds on a carpet, a wall coating, and a gypsum board in a test chamber. Indoor Air 3:276–282. https://doi.org/10.1111/j.1600-0668.1993.00009.x
16. De Bortoli M, Knoppel H, Columbo A Kefalopoulos S (1996) Attempting to characterize the sink effect in a small stainless steel test chamber. In: ASTM STP1287, pp 305–318
17. Dunn JE, Chen T (1993) Critical evaluation of the diffusion hypothesis in the theory of porous media volatile organic compounds (VOC) sources and sinks. In: ASTM STP 1205, pp 64–80 (1993)
18. Norhayati M, Hazim A (2010) The spatial distribution of carbon dioxide in an environmental test chamber. Build Environ Ventilation for Better Productivity, Comfort and Health (Roomvent Conference 2009) 45(9):1993–2001
19. Bai X (2004) Multi-dimensional CFD simulation of adsorption/desorption processes in carbon canister. Doctoral Thesis, Mechanical and Aerospace Engineering, University of Missouri, Rolla, United States
20. Bjorn E, Nielsen PV (2002) Dispersal of exhaled air and personal exposure in displacement ventilated rooms. Indoor Air 12(3):147–164. https://doi.org/10.1034/j.1600-0668.2002.08126.x
21. Gao N, Niu J (2006) Transient CFD simulation of the respiration process and inter-person exposure assessment. Build Environ 41:1214–1222. https://doi.org/10.1016/j.buildenv.2005.05.014
22. Shi B, Elsaved RK, Dadah A, Mahmoud S (2014) Simulation of honeycomb adsorption bed for automotive cooling system. In: Proceedings of the international conference on heat transfer and fluid flow Prague, Czech Republic, August 11–12, Paper No 180
23. Wu YX, Wang X, Ching CB (2002) Computational fluid dynamics simulation of the adsorption separation of three components in high performance liquid chromatography, Originals Column Liquid Chromatography. Chromatographia 55(7):439–445
24. Finch BSC (2011) Modeling transport and protein adsorption in microfluidic systems. Doctoral thesis. College of Sciences, University of Central Florida, Orlando, Florida, United States
25. Chinthala S, Gulia S, Silva D, Rana C, Mower K, Dengel A, Khare M (2016) Emission characteristics of eco-friendly sustainable building construction material-Part II. In: Indoor air conference, Ghent, Belgium
26. Gulia S, Chinthala S, Silva D, Rana C, Mower K, Dengel A, Khare M (2016) Emission characteristics of eco-friendly sustainable building construction material-Part I. In: Indoor air conference, Ghent, Belgium

Hospital Indoor Air Quality in Respect to Transmission of Infection

Prabir Kumar Sen and Parijat Sen

Abstract Airborne organisms, common in the hospital environment, can pose serious threats to patients—immune-suppressed and immune-deficient patients in particular. Though many of the infections in hospitals are transmitted through hand contact, surgical appliances, catheterization, intubation, or while on ventilation, it is an accepted fact that most of the opportunistic pathogens causing hospital-acquired infections (HAI) are at least partly airborne. They may be non-respiratory, but they get partly airborne before settling on the wound, or medical equipment/appliances. Setting in proper indoor air quality (IAQ) parameters, like temperature, humidity, dilution, filtration, pressurization, properly locating air terminal units, supported by planned installation, operation and maintenance through a robust protocol, reduces growth, count and transportability of infectious pathogens in hospital environment. Various codes and standards of American Society for Heating Refrigeration and Air Conditioning Engineers (ASHRAE), World Health Orhanisation (WHO), National Building code (NBC) and Facility Guideline Institute (FGI) give guidance in regard to IAQ in hospitals. A snapshot of healthcare industry shows that many hospitals are not following the guidelines. Advanced infection control measures, like ultraviolet germicidal irradiation (UVGI) and photocatalytic oxidation (PCO), though being used, are not being located properly. Even IAQ is not being mentioned as an important parameter in the infection control manuals for hospitals. This is causing large number of hospital-acquired infection and associated deaths.

Keywords Healthcare · Indoor air quality · Infection control · Current status

P. K. Sen (✉)
Consulting Engineer—HVAC, Kolkata, India
e-mail: prabirkumarsen@yahoo.co.in

P. Sen
Department of Pulmonary, Critical Care and Sleep Medicine, University of Kentucky, Lexington, KY, USA

A. Sharma et al. (eds.), *Indoor Environmental Quality*,
Lecture Notes in Civil Engineering 60,
https://doi.org/10.1007/978-981-15-1334-3_7

1 Introduction

Healthcare facilities house together individuals with infections and individuals susceptible to infections in close proximity, risking transmission of infections from one to another. It is not just patients, but even hospital workers, who are at risk of contracting infections in hospital.

A patient, during his stay in a hospital, may pick up infection. This is known as nosocomial infection. It is an infection which was not present, or incubated in the person, when he was admitted in hospital. He acquired it during his hospitalization. Being also known as healthcare-associated infection (HAI), or hospital-acquired infection, it can be caused by viral, bacterial or fungal pathogens. Typically, it manifests within 48 h of hospital admission to 30 days after discharge from the hospital.

Outpatient clinics and the patient waiting areas may have undiagnosed infected patients. However, these areas are generally air-conditioned similar to standard office areas. Emergency rooms (ER) are also at risk, as often people with undiagnosed infectious diseases may be taken in these rooms right away without detailed evaluation.

Hospitals often deal with immuno-compromised patients, comprising of both immuno-deficient and immuno-suppressed types. Failure or weakening of normal functioning of one or more elements of one's immune system may make him immuno-deficient. This includes patients with genetic immuno-deficient conditions, patients with infections like HIV, as well as patients undergoing chemotherapy. Individuals with autoimmune diseases, or those undergoing organ transplants, are often immune-suppressed deliberately with medications. These immuno-compromised patients are more susceptible to healthcare-associated infections.

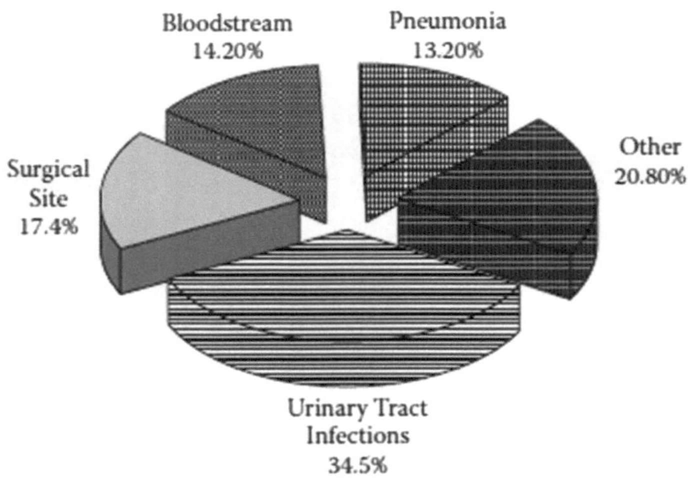

Fig. 1 Types of nosocomial infections [1]

The most common types of healthcare-associated infections (Fig. 1) are bloodstream infection (BSI), pneumonia (including ventilator-associated pneumonia, or VAP), urinary tract infection (UTI) and surgical site infection (SSI).

Prevention of healthcare-associated infections has received much attention in recent years as they create a large health risk and economic burden on individuals, as well as the society. A good and sustainable indoor air quality (IAQ) has a great role in preventing transmission of infections in a hospital. A large number of healthcare facilities, all over the world, are still away from meeting the heating ventilation and air conditioning (HVAC) requirements for infection control. Interdisciplinary coordination, between HVAC engineers and healthcare personnel, is a key to raising awareness on these critical issues, and tackling healthcare-associated infections.

2 Fact Sheet

WHO fact sheet [2] states that, of every 100 hospitalized patients at any given time, 7 in developed countries and 10 in developing countries acquire at least one healthcare-associated infection. Centre for Disease Control (CDC), USA [3], estimates that there are 722,000 HAIs in acute care hospitals and 75,000 associated deaths in the USA each year. Unfortunately, not much organized data on occurrence of healthcare-associated infections is available for low- and middle-income countries (LMIC).

HAIs have large prevalence in intensive care units (ICUs) and in acute surgical and orthopaedic wards. In high-income countries, approximately 30% of the patients [3] in intensive care units (ICU) are affected by at least one healthcare-associated infection. In low- and middle-income countries (LMIC), the frequency of ICU-acquired infections are about 2–3 folds higher. Device-associated infections are also up to 13 times higher [3] than that in the developed countries. In ICUs, a third of the healthcare-associated infections are respiratory type.

3 Mechanism of Infection Transmission

Infection occurs when all of the following elements of its transmission chain are present in a healthcare environment.

- An infectious agent, i.e. an infectious micro-organism.
- A source (of the agent), i.e. a patient with infection.
- A susceptible host, i.e. an immune-compromised (including immune-weakened, immune-deficient and immune-suppressed) person, to receive the agent.
- And most critically, a pathway for the agent to travel from the source to the host.

Generally, the severity of the impact of infection on a person depends on the length of exposure to infection, virulence of the microbes to which he is exposed, location of the exposure and the infection load.

The infections may take following routes.

- Contact—through touch, stethoscope, bedpans, catheters, dressing, gloves, needles, appliances, instruments, ventilators, etc.
- Aerial route/airborne—through droplets and secretions on surface and air, inhalation of infectious particles, etc.
- Oral route.
- Vectorborne through flies, mosquitoes, rats, etc.

4 Role of Indoor Air Quality in Infection Control

Some studies indicate that only 10% healthcare-associated infections is airborne, while others state that 16% HAIs in ICUs is airborne. The fact still remains that more than a third of HAI agents possibly get airborne at some part of their transmission. Though many of the infections are transmitted through contact, more so on sites of invasive procedures (like surgery, catheterization, intubation or putting on ventilation), it is possible that the infectious microbes have taken an aerial path at some stage of their travel. Many opportunistic pathogens though non-respiratory, get partly airborne before settling on the wound or medical equipment/appliances.

Formation of aerosols from contaminated water, and their getting airborne, is also a major mode of spread of infection, like Legionella, causing sporadic cases of Legionellosis (a healthcare-associated pneumonia).

Given the above nature of transmission of infections, indoor air quality (IAQ) has a major role in the healthcare industry, particularly in combating spread of infections. Among other measures, proper engineering design and planning of HVAC systems, supported by selection, of proper technology, equipment, material, control, air intake/exhaust locations, appropriate protocols, and periodic monitoring and maintenance can contribute greatly to reduce healthcare-associated infections. Unfortunately, most healthcare facilities, in low- and medium-income developing countries in particular, do not identify this as a critical aspect of infection control.

5 Control of IAQ Parameters for Control of Infection

Infection control science is a complex blend of engineering, particle physics, microbiology and medicine. Good HVAC systems can help to control infections by controlling different IAQ parameters, like temperature and humidity (reducing growth), dilution (reducing exposure time), filtration (reducing count), properly locating supply air

Air Changes per Hour, ach	Time Required for Removal Efficiency of 99%, min	Time Required for Removal Efficiency of 99.9%, min
2	138	207
4	69	104
6	46	69
8	35	52
10	28	41
12	23	35
15	18	28
20	14	21
50	6	8

Fig. 2 Effect of air change rates on particle removal [4]

terminal units (preventing patient discomfort and coughing) and controlling airflow patterns by suitable pressure gradient (reducing induction and transportability).

Temperature and humidity control not only aid in wound healing and patient comfort, but also prevent growth of potential infectious agents. Patients may also start coughing/sneezing in low relative humidity (RH) situation. Location of supply air terminals is equally important, as a direct blast of cold air on patients' body may prompt them to cough/sneeze and release innumerable droplets, which may have infectious microbes that may travel a long distance in air stream, particularly when the air flow is turbulent.

Rate of dilution, of room air by outdoor air, controls the number of micro-organisms present in a room and their exposure time on individuals. ASHRAE HVAC Design Manual for Hospitals and Clinics [4] presents a table (Fig. 2) relating room air change rates and particle removal time. It indicates that it will take 23 min to remove 99% of the particles from a room with a dilution rate of 12 air changes per hour.

Some researchers suggest that only about 7% of the hospital area needs air filtration with MERV 14, or better class of filters. This may be an understatement considering the submicron size of bacteria, complexity of patient mix and space crunch in present day hospitals, where a large number of facilities are packed in one building in a congested area. Orthopaedic surgery and bone marrow transplant rooms should use HEPA filters (MERV 17) to keep their submicron bacteria level low. Protective environment (PE) rooms for immune-compromised patients should also use HEPA filters. General operating rooms, ICUs, recovery rooms, and in fact, in all in-patient areas, diagnostic and treatment areas should use MERV 14 filters to keep the bacteria level low. Even the exhaust air from negative pressure isolation rooms for contagious patients should pass through HEPA filters before being released in the congested surrounding.

Use of other additional advanced devices of air purification can enhance indoor air quality further. Ultraviolet (UV) radiation inactivates micro-organisms by damaging their DNA and RNA strands (Fig. 3) by penetrating their cell walls, and making

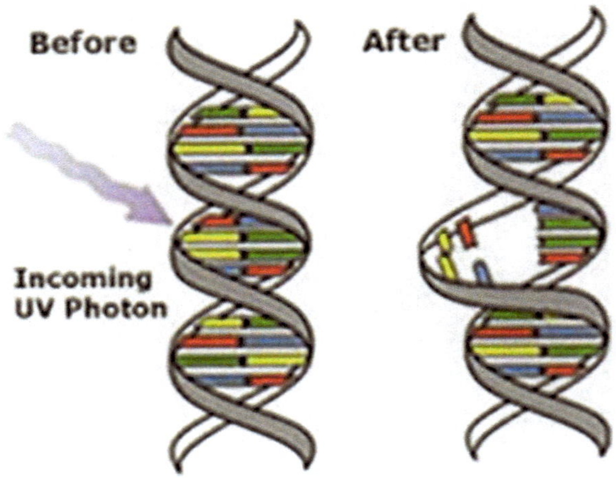

Fig. 3 UV ray inactivates micro-organisms [5]

them unable to replicate and propagate infections. This can effectively reduce the virulence of micro-organisms and, therefore, can reduce infection rates.

Photocatalytic oxidation (PCO) units are air purification devices, next generation to ultraviolet germicidal irradiation (UVGI) units. It uses broad-spectrum UV light to impinge (Fig. 4) on a thin film of titanium dioxide, to create free hydroxyl (OH^+) radicals that can neutralize micro-organisms. It can eliminate particles as small as 0.001 microns, compared to 0.3 microns with HEPA filters.

Both UVGI and PCO units are especially effective while used on static particles. They can keep cooling coils, filters and drain pans of air handling units free of microbial colony. Efficacy of these devices are yet to be ascertained while acting on fast-moving micro-organisms, like the ones in air inside the ducting, or in room air.

The fast-moving micro-organisms get much less exposure of UV/free radicals. The UV/free radical dose has to be high in such event.

Some areas in a hospital are maintained at positive or negative pressure (with respect to surrounding areas) to control transmission of infection. Operating rooms

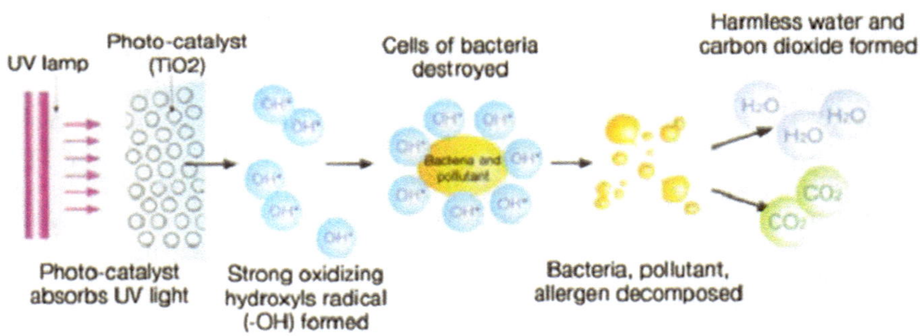

Fig. 4 Working of photocatalytic oxidation unit [5]

and rooms for immune-compromised patients are maintained at a positive pressure in contrast to airborne infection isolation (AII) rooms, which are kept at a negative pressure. Objective of pressurization is to direct potentially infectious micro-organisms from the more sterile to less sterile areas. An anteroom is suggested with pressurized rooms to prevent sudden change in room pressure. The following areas in a hospital are recommended to be positively pressurized as per Facility Guidelines Institute (FGI) guidelines/standard 170-2013.

- Operating rooms
- Delivery rooms
- Trauma rooms
- Newborn intensive care (NICU)
- Laser eye surgery rooms
- Protective environment rooms for immuno-compromised patients
- Pharmacy
- Laboratory, media transfer
- Central Medical and Surgical Supply Clean workrooms
- Central Medical and Surgical Supply Sterile Storage.

Following are the areas in a hospital that should be maintained at negative pressure as per FGI guidelines.

- ER waiting rooms
- Radiology waiting rooms
- Triage
- Toilet rooms
- Airborne infection isolation (AII) rooms
- Darkrooms
- Cytology, glass washing, histology, microbiology, nuclear medicine, pathology and sterilizing laboratories
- Autopsy rooms
- Soiled workrooms or holding rooms
- Soiled or decontamination room for central medical and surgical supply
- Soiled linen and trash chute rooms
- Janitors' closets.

6 Sample Survey

A sample survey was carried out in a metropolitan city in India to get an idea of the current status of IAQ control in healthcare facilities. Two corporate multidisciplinary hospitals, three Govt. tertiary care centres and three small healthcare facilities were visited in September, 2017 with an objective to study their HVAC systems and IAQ control measures in relevance to controlling healthcare-associated infections. Emphasis was on HVAC systems for operation rooms and intensive care units, the two major areas for propagation of healthcare-associated infections. The observations vis-a-vis recommendations of various codes/standards/guidelines are presented in Table 1.

7 Discussion

Air conditioning, in many healthcare facilities, more so in low and medium income countries (LMIC), is still considered as just a comfort measure for the rich patients. Certain protocol imposed hygiene control measures (like taking off shoes, floor mopping and using sterile equipment) are in practice. Maintenance of hand hygiene is also gaining importance. IAQ control, on the other hand, is hardly considered a measure for controlling healthcare-associated infection. It is considered, somehow, important only in controlling airborne respiratory infections. Relationship between IAQ and healthcare-associated infections in general is not given its due importance while designing, operating and maintaining HVAC system of a healthcare facility. This is quite evident from the observations of the small sample survey. Even some operation manuals for hospital infection control do not mention the importance of indoor air quality. Emphasis seems to be more on treating the patients with HAI, instead of preventing healthcare-associated infections, a very important root cause of which is IAQ.

Cost is a major constraint in implementing the desired level of indoor air quality in a healthcare set-up, particularly in LMICs. However, what is often missed is the final burden that arises out of ignoring this aspect. A few of these issues, such as spacing the exhaust and outdoor air intake and keeping the AHU rooms free from unwanted materials, can be taken care through awareness and planning, and at no additional cost. Stake holders should also understand that the benefit of 100% exhaust from an OR is lost once they place its outdoor air intake in close proximity of the OR exhaust duct. It was seen in the survey that using AHU rooms as a free storage space is a popular practice. This increases the pollution load in critical areas of the hospital.

When HEPA/MERV 14 filters are used for a positive pressure OR, or a protective environment room, it is always better to continuously monitor (Fig. 5) its pressure levels with alarm systems to alert in case of any eventual depressurization making a pathway for the micro-organisms.

Standard commercial type package AC units generally do not have high static pressure evaporator fans. Using MERV 14 filters in supply air duct with them will compromise on the supply air rate. This may lead to micro-organisms staying longer in the room, increasing the chances of infection.

In the survey, the ICUs were found to be a neglected area as far as the control of IAQ is concerned. NBC of India and ASHRAE guidelines should be followed for outdoor air and supply air change rates in ICUs. Recirculating type residential/commercial air conditioners (Fig. 6) are being used largely in the ICUs, though they do not meet any infection control requirement (outdoor air, filtration and ACPH rates). Curtain type partitions may provide some visual privacy with flexibility, but using such partitions

Table 1 Observations vis-a-vis recommendations of various codes/standards/guidelines

	Feature	Observation	Recommendations
1	Air change per hour in operation rooms and ICUs	Some ORs are using 40% outdoor air, some are using 60% and some are using 100% outdoor air for dilution. In one hospital (using 100% outdoor air) from the sample, outdoor air intake for OR is just adjacent to OR exhaust duct, causing recirculation. Most ICUs in smaller healthcare facilities, and about 16% of ORs in the sample are using commercial packaged air conditioners rendering no concern about outdoor and supply air change per hour (ACPH) rates. Even residential/commercial recirculating type hi-wall split air conditioners/cassette type units (Fig. 6) with zero outdoor air are quite popular in ICUs of smaller healthcare facilities.	National Building Code (NBC) of India [6] suggests that recirculation of air should be avoided in ORs. Outdoor air with minimum 5 air changes per hour (ACPH) and room supply air with 25 ACPH shall be used. ICUs and recovery rooms shall have 6 ACPH of room supply air and 2 ACPH of outdoor air. ASHRAE 170 suggests minimum outdoor air at 4 ACPH, room supply air at 20 ACPH for ORs, and 6 ACPH of room supply and 2 ACPH of outdoor air for ICUs. Both ASHRAE 170 and NBC of India suggest that air exhaust and intake points should be separated by at least 8 m distance. ASHRAE HVAC Design Manual for Hospitals and Clinics [4] says that use of standard recirculating room HVAC units are not accepted for ICUs.
2	Air filtration in operation rooms and ICUs	37% of the hospitals from the sample are using HEPA Filters (MERV 17) in ORs, while others are using duct mounted fine filters (MERV 14). Some are using duct mounted MERV 14 filters with standard package air conditioners. Many ICUs, PICUs, NICUs and even some ORs are using packaged air conditioners with MERV 7/MERV 14 filters, though the evaporator fans of these units do not develop adequate static pressure required for use with MERV 14 filters in supply air duct. In some cases, common PAC units are being used for OR, Recovery Room, ICU and patient ward.	ASHRAE HVAC Design Manual for Hospitals and Clinics and NBC of India suggest MERV 14 filters for general ORs and ICUs. They suggest HEPA filters in orthopaedic ORs and bone marrow/organ transplant rooms and also in pressurized protective environment (PE) rooms. The Facility Guidelines Institute (FGI) suggests a filtration level of MERV 14 for all patient care areas, whether in clinics or in full-service hospitals.

(continued)

Table 1 (continued)

	Feature	Observation	Recommendations
3	Room air supply system.	Many ICUs, and even ORs, are using continuous grille (Fig. 7) for inline supply and return air. Locations of supply air grilles in many ICUs are right over/near the patient bed.	NBC of India recommends that supply air outlets shall be located at, or near the ceiling, and return/exhaust be collected near the floor level to ensure clean conditioned air to move through the patient/working space before it collects room particles and moves to return/exhaust at floor level. ASHRAE HVAC Design Manual for Hospitals and Clinics says that patients can be especially sensitive to cold air currents from a supply air diffuser
4	Pressurization	Most ORs have been designed to maintain '+ve' pressure. But over-pressure monitoring device (Fig. 5) is present only in 25% of the places. Most ICUs are having dormitory type layout with no positive/negative pressure rooms. There are some curtain partitioned spaces (Fig. 6) in some of these open plan ICUs. A few of them have separate once through supply air system to use them as '−ve' pressure rooms. Some ICU floors have a few rigid partitioned '+ve' and '−ve' pressure rooms. But the return air (RA) from these rooms are taken in a common duct and re-circulated through a common air handling unit (AHU), which is also catering to the adjoining dormitory type ICU floor.	ASHRAE 170 suggests both ORs and ICUs shall have positive pressure, whereas NBC of India suggests positive pressure for ORs and neutral pressure for ICUs and recovery rooms. ASHRAE HVAC Design Manual for Hospitals and Clinics suggests continuous monitoring of pressurization in '+ve'/'-ve' pressure rooms. NBC of India, however, suggests pressure verification semi-annually. There is no guideline on whether, or not, ICUs shall have rigid partition. A number of professional and scientific bodies now emphasize the importance of rigid isolation facilities in ICUs—at least one cubicle for high-risk patients per eight ICU beds.
5	Ducted return air	Return air is not ducted in many ORs and ICUs. Space above false ceiling is used as return air (RA) plenum. In some cases, even after ducting the room return air to AHU rooms, the RA is left open in the AHU rooms, which have free access	NBC of India suggests that both supply air and return air shall be ducted for critical areas. NBC of India also suggests that access to equipment rooms shall be controlled.

(continued)

Table 1 (continued)

	Feature	Observation	Recommendations
6	Use of ultraviolet germicidal irradiation (UVGI) and photocatalytic oxidation (PCO) units	UVGI and PCO units are being used in some ICUs, HDUs and ORs. They are used over AHU cooling coils/SA grilles/RA duct, or at upper level of OR and ICU floors.	NBC of India suggests their use on evaporator coil, either on its upstream or downstream side.
7	Other observations	Only a few hospitals conduct periodic particle count on indoor air of ORs and ICUs to assess bacteria level.	NBC of India suggests that testing, adjusting, balancing (TAB) and evaluation of IAQ shall be performed and recorded once every year.
		Fumigation of ORs and ICUs are rarely done.	National Accreditation Board for Hospitals & Healthcare Providers (NABH), India, suggests fumigation in the high-risk areas like ICU, PICU, NICU, labour room and OT.
		Most AHU rooms are used as storage space for redundant material/filters/components, installation/maintenance tools and even housekeeping accessories, like mops, brooms, spades and buckets. Even a data server has been installed inside the AHU room of PICU and NICU of the paediatric centre of a hospital.	NBC of India cautions that air handling unit rooms shall not be used as a storage space for storing files and waste materials.
		Outdoor air intake, in one case, has been found to be close to cooling tower.	ASHRAE 170 suggests that outdoor air intakes for air handling units shall be located at least 8 m away from cooling towers and all exhaust and vent discharges.
		Maintenance work over false ceiling was found being done in one working ICU without any sealing of the work area.	NBC of India suggests that this should be done by sealing off the areas to make certain that airborne maintenance debris is unable to get into the air.
		Soiled scrubs are dumped in one corner of the corridor of OR and recovery room area	ASHRAE 170 suggests that soiled utilities shall be kept in a negative pressure room with 100% exhaust.

Fig. 5 Room pressurization monitor outside door (circled red)

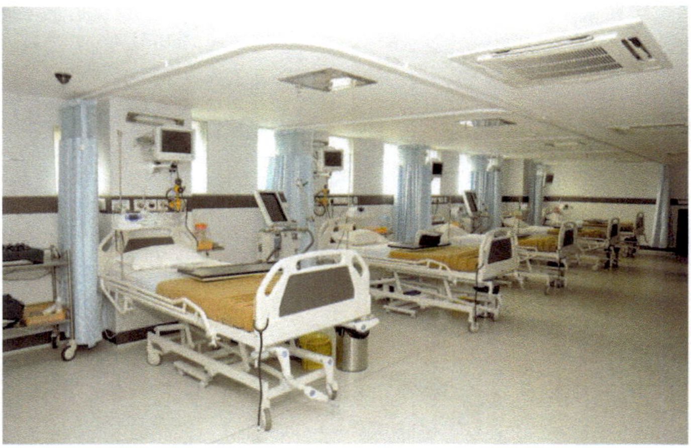

Fig. 6 Curtain partitioned ICUs with commercial air conditioners

to build a negative pressure isolation room cannot eliminate chances of propagation of infection within the ICU complex. Though there is no definite recommendation in place, use of rigid partitioned cubicles in ICUs are getting popular. This may be costly, but is definitely a step forward towards infection control in ICUs. Positive pressure and negative pressure rooms should have solid partitions, and be equipped with separate individual air systems, instead of running it through a common air handling unit. In case, a single room is used as both '−'ve and '+'ve pressure rooms, it should be properly disinfected before switching its use from '−'ve pressure to '+'ve pressure application.

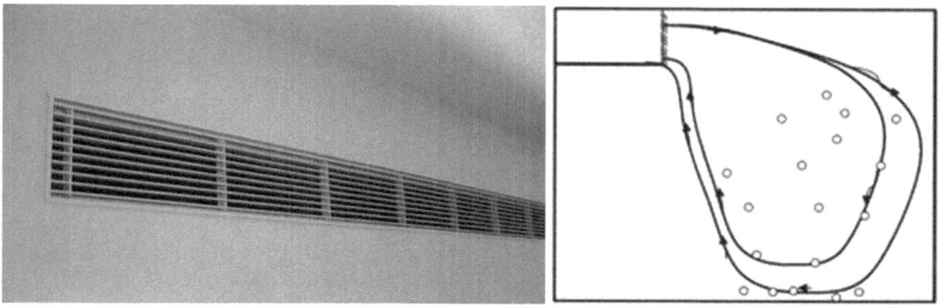

Fig. 7 Continuous supply and return air grille in an ICU

Location of supply air grilles over a patient's bed may make patients cough and sneeze, and release infectious droplets in the air. Using continuous grilles for inline supply and return air will mix up supply air with all the pathogens in the room (Fig. 7) before it reaches the patient level. Hence, supply air should be from the top and return air from bottom as suggested by the codes.

Using the space above false ceilings as return air plenum may induct moisture in this negative pressure space from the surrounding areas. This may lead to damp patches leading to growth of moulds and spores, which may finally get into the room air system. Both supply and return air in hospitals should be ducted. The ducting should be checked for leaks and connected directly to air handling units instead of leaving return air open in the AHU rooms.

UVGI and PCO devices maybe used to irradiate static parts, like coils, filters and drain pans, which normally collect the room micro-organisms. Trying to target micro-organisms in airborne condition in ducting, or room air by UVGI/PCO units is unlikely to be effective. Lockout switches should be used to switch off UVGI/PCO units before accessing the air handling units.

8 Conclusion

Healthcare-associated infections are major concern for healthcare facilities in any country. The threat has got even bigger with the emergence of drug-resistant infection.

Though new standards, protocols and thoughts are coming in to mitigate the problem of healthcare-associated infections, a large number of healthcare facilities (particularly the old ones) all over the world are lagging behind. There is a need for their review and upgradation in light of current norms.

Prevention of HAIs is possible, but it will need a conscious effort from all individuals including clinicians, healthcare administrators, system (including HVAC) teams, public health officials, quality improvement groups and the government bodies. All should work together towards improving hospital indoor air quality, which in the long run will serve in protecting the patients.

Finally, formation of a regulatory body for monitoring and regulating HVAC systems and IAQ standards in healthcare system will help in streamlining this effort.

References

1. Kowalski W (2007) Air-treatment systems for controlling hospital-acquired infections. HPAC Engineering 79
2. World Health Organisation—Healthcare associated infections FACT SHEET
3. Centre for Disease Control (CDC), USA—https://www.cdc.gov/hai/surveillance/index.html
4. ASHRAE HVAC Design Manual for Hospitals and Clinics, Second edition
5. Product brochure, Aeropure UV System Pvt. Ltd. Pune, India
6. National Building Code of India (2016)
7. Prevention of Hospital–Acquired Infection, A Practical Guide (2nd Edition, 2002), World Health Organisation
8. ICU Planning and Design in India—Guidelines 2010, Guideline Committee, Indian Society of Critical Care Medicine

Environmental Monitoring of $PM_{2.5}$ and CO_2 in Indoor Office Spaces of Delhi, India

Akansha Gupta, Radha Goyal, Priyanka Kulshreshtha and Ashish Jain

Abstract Delhi ranks highest among the most polluted city in the world in terms of air pollution. Its health impact may include diseases like asthma, lung cancer, COPD, increased long-term risk of cardiopulmonary mortality. Degraded indoor air quality inside commercial buildings such as offices may affect the health of the workers and can indirectly affect their productivity. In the present study, indoor air quality has been studied in four different air-conditioned office buildings located in Delhi NCR for the criteria pollutant $PM_{2.5}$ and the CO_2 as ventilation parameter. The total hazard ratio indicator has also been calculated from the data of $PM_{2.5}$ and CO_2 for all selected office premises. The results of the study show the highest concentration of $PM_{2.5}$ in building A1 (116.5 ± 67 $\mu g/m^3$) and highest CO_2 concentration in building A2 ($1600 + 30.5$ ppm). Higher concentration of $PM_{2.5}$ in building A1 could be due to its maximum proximity to urban busy roads and poorly maintained HVAC ducting system, which may lead to infiltration and more leakages of $PM_{2.5}$ from outdoors. Similarly, the highest concentration of CO_2 in building A2 could be due to insufficient ventilation condition. In each studied building, the concentration of CO_2 and $PM_{2.5}$ are recorded to be higher than the NAAQS and ASHRAE standards. The health hazard ratio indicates that both CO_2 and $PM_{2.5}$ plays an important role in determining the health of the building. However, the health impacts of increased $PM_{2.5}$ could be more severe than CO_2 depending upon the sources of $PM_{2.5}$.

Keywords $PM_{2.5}$ · CO_2 · Offices · Indoor air quality · Total hazard ratio

1 Introduction

In 2014, World Health Organization (WHO) recognized air pollution exposure as the single largest environmental health risk which causes one-eighth of total deaths in 2012. Currently, around 80% of the total population living in India relies on combustion of biomass fuels for cooking purpose [1]. Many people link air pollution, only

A. Gupta (✉) · R. Goyal · P. Kulshreshtha · A. Jain
Indian Pollution Control Association, New Delhi, India
e-mail: akanshagupta05.ag@gmail.com

A. Sharma et al. (eds.), *Indoor Environmental Quality*,
Lecture Notes in Civil Engineering 60,
https://doi.org/10.1007/978-981-15-1334-3_8

with the outdoor pollution, however, high concentrations of pollutants, commonly occur in the indoor environment. According to the United States Environmental Protection Agency (USEPA) [1] in urban environment, the concentration of indoor air pollutants is much higher than the outdoor pollutants which affect the quality of life and the well-being of the population in general. People spend more than 70% of their time indoors such as inside residential houses, school, college, commercial or industrial buildings [2, 3]. It is ranked among the top five environmental health risks to the public by USEPA. In India, indoor air pollution is one of the top 10 deaths, disease risk factors according to WHO. Delhi being a metropolitan city of India ranks highest in terms of air pollution which has caused an alarm for people and health experts. It is estimated that in Delhi 3000 metric tons of air pollutants are emitted everyday in which the major contribution is from vehicular emissions (67%) and emission from coal-based thermal power plants which hold 12% [4].

Poor indoor air quality (IAQ) can be harmful to susceptible groups such as children, elderly, and those with chronic respiratory disease and cardiovascular like asthma. Apart from its profound effect on health, the poor indoor air quality reduces the comfort and productivity of occupants of the building. In commercial buildings, such as offices, poor IAQ may affect the health of the employees as well as indirectly affect their productivity. The researches on IAQ in offices are very limited in comparison to the magnitude of the population in commercial buildings. Most published literature in India has focused on indoor air pollution caused by biomass fuel burning from traditional cookstoves. Many indoor pollutants are responsible for indoor air pollution such as poor ventilation system, prevalent use of tobacco smoke, electronics printers, furniture's, paints, and computers in urban settings. Particulate matter of the size 2.5 $\mu g/m^3$ concentration in Delhi/NCR is the highest among many cities around the world [5] Exposure to $PM_{2.5}$ is associated with an increased long-term risk of cardiopulmonary mortality by 6–13% and is also associated with a reduction in life expectancy up to 20 months [6]. Due to inefficient air circulation in the indoor environment, most of the outdoor pollutant enters inside such as building materials (cement, wood, etc.), industrial and vehicular exhaust, etc. [7].

Ventilation plays an important factor in determining the indoor air quality in a building, to minimize the energy cost, and minimizing the outdoor air used for ventilation is a common practice adopted by buildings. This lead to compromising the health and comfort of the occupants [8] CO_2 concentration inside is an important factor in determining the IAQ with respect to ventilation. CO_2 has been majestically associated with a variety of outcomes including an increase in sick leaves at an office [9], symptoms of influenza and sick building syndrome. It may also adversely affect human cognition and decision-making performance [10]. A study was conducted by Allen et al. where participants were exposed to 550, 945, and 1400 ppm concentration of CO_2 for 8 h and found that cognitive function scores were 15 and 50% lower from the concentration level at 550 ppm. This indicates that effects of CO_2 on human cognitive performance.

Thus, the indoor of the office buildings become the subject of much attention. Various studies had been conducted on the indoor air quality in office buildings from different parts of the world [11, 12]. However, India has few studies related to

IAQ in offices [13–15]. In this study, continuous (8 h/day), $PM_{2.5}$ and CO_2 levels in different types of an air-conditioned office building located in Delhi/NCR during the winter season were measured. To compare the health effects of the different indoor environment, integrated IAQ total hazardous ratio indicator (THRI) was also calculated. This study aims at analyzing and comparing the indoor air quality in commercial buildings of central Delhi with its impact on health and productivity.

2 Methodology

2.1 Site Description

The present study aimed at assessing the indoor air quality in four office buildings (A1, A2, A3, and A4) located in the commercial areas of Delhi/NCR (National Capital Region). The study was conducted during the month of January and February 2018, which was a critical winter period with mean ambient temperature and humidity of 14 °C and 79%, respectively.

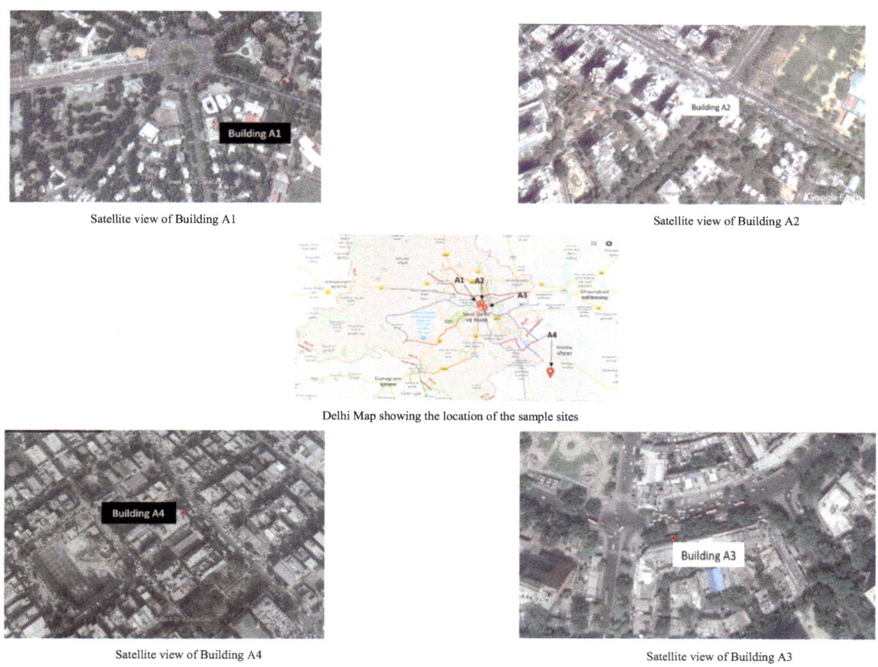

Fig. 1 Delhi map showing the location of the sampling sites

Table 1 Building characters of the selected buildings

Characteristics	A1	A2	A3	A4
Age of building	50 years	20 years	35 years	10 years
Type of flooring	Brick-walled with marble flooring	Brick and concrete with marble and carpet flooring	Brick-walled with marble flooring	Brick and concrete with marble and carpet flooring
Location of the building	The central part of the Delhi, two sides facing the busy roads	Central part of the Delhi, one side of the building is facing busy traffic	Central part of the Delhi, two sides of the building were facing the roads	Industrial area of Noida
Number of floors	4	15	4	12
Type of HVAC system	Centrally air-conditioned	Centrally air-conditioned	Split and cassette two-way air-conditioning units	Centrally air-conditioned
Condition of HVAC system	The ducting system at building A1 was very old and was not cleaned properly. The building was negatively pressurized	Fairly maintained ducting system. The building was negatively pressurized	Open gallery in between, which was the source of fresh air inside the building. The building was negatively pressurized	Fairly maintained ducting system. The building was negatively pressurized
Building activity	Regular office activities	Regular office activities	Regular office activities	Regular office activities
Occupancy	34	45	56	123

Three office buildings (A1, A2, and A3) were located in central part of Delhi and one (A4) in the Noida Region of NCR (Fig. 1). During the monitoring period, the ambient air quality index [16] in all the locations was under unhealthy (Table 1).

3 Building Characteristics

3.1 Monitoring and Analysis

The IAQ sampling/monitoring of all four selected buildings had been conducted during regular office hours (8 h) in weekdays. The indoor and outdoor concentration

Table 2 Average concentration of $PM_{2.5}$ and CO_2 at each location

Site	Location	Temp. (°C)	rH (%)	CO_2 (ppm)	$PM_{2.5}$ indoor (μg/m^3)	$PM_{2.5}$ outdoor (μg/m^3)	I/O Ratio
A1	Ground floor	24.8 ± 1	44.8 ± 1	1578 ± 273	124 ± 26	187.53	0.66
	First floor	22 ± 2	48 + 6 ± 1	1013 ± 197	109 ± 49	187.53	0.58
A2	7th floor	24.6 ± 2	44.7 ± 3	1443.7 ± 968	74.6 ± 12	187.53	0.40
	11th floor	24.8 ± 1	47 ± 2	1758 ± 365	73.6 ± 49	187.53	0.39
A3	Ground floor	23.2 ± 1.5	49.6 ± 5	894 ± 159	53.3 ± 56	95.3	0.28
A4	Ground floor	25.8 ± 1	50.6 ± 3	1918 ± 298	76 ± 31	76.0	1.07
Prescribed limit		23–26[b]	30–70[b]	1000[b]	60[a]	60[a]	

[a]$PM_{2.5}$ standard from National Ambient Air Quality Standard (NAAQS)-2011
[b]CO_2 standard is from American National Standard Institute (ANSI)/ American Society for Heating, Refrigeration and Air conditioning Engineers (ASHRAE) Standard 62.1-2007

of $PM_{2.5}$ and CO_2 in each building was monitored at every 5 min interval by pre-calibrated DustTrak Aerosol Monitor II–Model-8533 (TSI, USA) and Testo IAQ Monitor 435-2 (Testo, Vienna, Austria), respectively. The DustTrak is a real-time optical scattering instrument that measures $PM_{2.5}$. The concentration range of DustTrak is 0.001 to 150 mg/m^3 with the resolution of +0.1% of reading with the flow rate of 3.0 L/min with +5% accuracy. The DustTrak was zero calibrated before starting of monitoring each day using zero calibrator instruments. The CO_2 measurement range of the Testo IAQ Monitor was 0 to +100,000 ppm with +75 ppm accuracy and 1.0 ppm resolution. Indoor humidity, temperature, and air pressure were also recorded. The 8-h average concentration along with standard deviation of all monitored parameters has been shown in Table 2 along with the average indoor–outdoor ratio which was also calculated for $PM_{2.5}$ for all buildings.

4 Total Hazard Ratio Indicator

The total hazard ratio was calculated for the recorded $PM_{2.5}$ and CO_2 concentration at each building and compared with reference values [17]. Mean reference concentration of $PM_{2.5}$ for this study was taken as 60 μg/m^3 (NAAQS, 2011); whereas, that of CO_2 was considered as 1000 ppm [18]. The hazard ratio for each air quality parameter, i.e., $PM_{2.5}$ and CO_2 (HR) were calculated by dividing the average concentration of selected air quality parameter, by its corresponding reference concentration (RfC), both expressed in the same unit:

$$\mathrm{HR}_i = C_i/\mathrm{RfC}_i$$

The total hazard ratio was also calculated to assess the global inhalation exposure risk of occupants in each studied building, for each building.

$$\mathrm{THR}_{\mathrm{site}} = \sum \mathrm{HR}_i$$

5 Results and Discussion

According to the ambient air quality index, the ambient air in the sampled locations was under highly or severely polluted category round the year. During the sampling period, the average temperature inside the offices on each floor was between 22 °C and 25 °C. The relative humidity was maintained between the limit, i.e., 30–60% according to the ASHRAE standard, 2013. All the four sampled locations were negatively pressurized; therefore, it can lead to contamination of indoor air with suspended particles or other particulate pollutants from the ambient air infiltration.

In all the sampled locations, windows were closed properly and there was no source of infiltration from them. However, in each building, there was no air curtain which made the main entry door is a major source of infiltration from the ambient air. The data collected was analyzed in depth to conclude the findings of the study which are discussed below.

6 Indoor $PM_{2.5}$ Concentration

In all the sampled location, the indoor concentration of $PM_{2.5}$ was recorded higher than the standards laid by NAAQS, i.e., 60 $\mu g/m^3$ due to higher ambient concentration. The average concentration of $PM_{2.5}$ at both ground (124 $\pm$ 26 $\mu g/m^3$) and first floor (109 $\pm$ 49 $\mu g/m^3$) of A1 building was recorded to be highest among all the buildings, i.e., 116 $\mu g/m^3$. This could be due to its highest proximity from busy roads and poorly maintained HVAC system. Further, the concentration at ground floor was higher than the first floor due to higher occupancy level and infiltration through the front and back door (entrance). The air-conditioning system was mostly switched off during the monitoring period, which may lead to lower circulation of air and accumulation of $PM_{2.5}$ levels. The $PM_{2.5}$ levels in buildings A2 (7th Floor), A2 (11th Floor), A3 and A4 are 74.6 $\pm$ 42 $\mu g/m^3$, 73.6 $\pm$ 49 $\mu g/m^3$, 53.3 $\pm$ 56 $\mu g/m^3$, and 76 $\pm$ 31 $\mu g/m^3$, respectively. Building A3 reported the lowest indoor $PM_{2.5}$ levels compare to others. This may be due to building design and ventilation system, which allows proper circulation of the air [24, 25]. The site also installed with exhaust fan, which helped to reduce the indoor $PM_{2.5}$ levels. In buildings, A2 and A4 indoor

$PM_{2.5}$ levels are comparatively higher. The observation datasheet indicates that the building A4 was having the highest occupancy and was fully carpeted, which may induce more indoor $PM_{2.5}$ levels.

7 Indoor CO_2 Concentration

The CO_2 concentration indoor is the surrogate index of ventilation in an occupied enclosed space. The ASHRAE recommends that indoor air CO_2 levels should be 1000 ppm (ambient + 645 ppm). The average CO_2 level in all the buildings except building A3 (894 ± 159 ppm) was exceeding the prescribed limit by ASHRAE (Table 1). It could be due to the passageway which was acting like a ventilation shaft and helping in maintain the CO_2 concentration exhaled by the occupants within the permissible limits. The highest CO_2 levels were reported in building A4 1918 ± 298 ppm due to the highest occupancy (123 employees) and no dedicated treated fresh air ventilation system installed in the building. The CO_2 concentration at A1 building was recorded 1578 ± 27 ppm and 1013 ± 197 1013 ppm, respectively at ground and first floor. In building A2, the CO_2 levels were recorded as 1443.7 ± 968 ppm at 7th floor and 1758 ± 365 ppm at 11th floor which makes the existing ventilation system inadequate for the present occupancy level [26, 27].

The CO_2 monitoring results also show that CO_2 concentration was continuously increased after post lunch period, i.e., after 2:00 PM (Fig. 2). This could be attributed because of the accumulation of the CO_2 concentration over the period of time, effect of food intake which leads to higher exhalation of CO_2 by the occupants [19] and

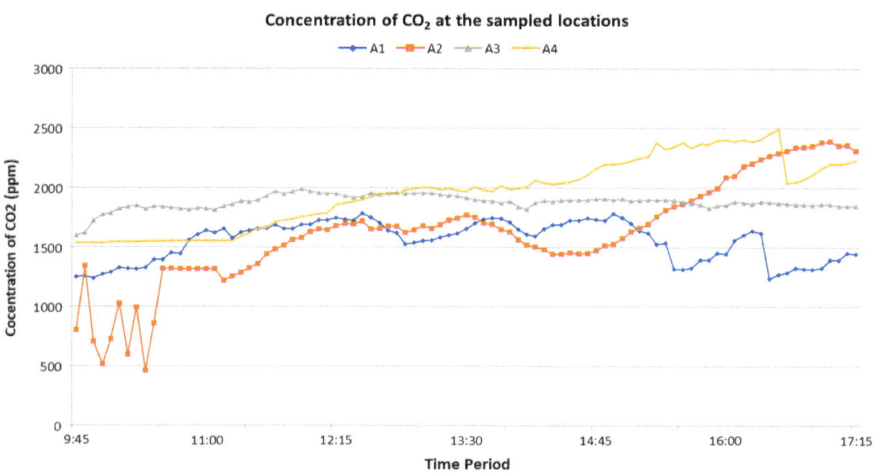

Fig. 2 Time scale graph of the CO_2 level for each sampled location

also the movement of visitors inside the buildings increases in the afternoon hours in comparison with morning hours.

8 Indoor/Outdoor Ratio Analysis

The I/O ratio for the $PM_{2.5}$ levels was found to be highest in building A4 (1.07), which indicates that there is indoor $PM_{2.5}$ sources exist in the building apart from infiltration from outdoors. The I/O ratio of building A1 of ground and first floor, A2 7th and 11th floor and A3 are 0.66, 0.58, 0.40, 0.39, and 0.28, respectively which indicated that the major source of increase in $PM_{2.5}$ concentration indoor was due to higher concentration at the ambient level.

9 Total Hazard Ratio Indicator (THRI)

With the reference concentration of $PM_{2.5}$ and CO_2 for daily exposure of 8 hours, the total hazard ratio of indoor air pollution was calculated at each building (Table 3). It was based on the non-carcinogenic risk assessment of the pollutant $PM_{2.5}$ and CO_2. The HR of $PM_{2.5}$ was highest at A1 building among all the buildings. However, the HR of CO_2 was significantly higher at the Y1 location of the A2 building. The Indoor THRI value was highest at A1 and A4 buildings. In building A1, highest THRI is due to highest $PM_{2.5}$ concentration, whereas in building A4, it is due to highest CO_2 levels. The calculated values of THRI at all study sites were much higher than earlier reported study [17, 20] in office buildings in Delhi and school in Italy.

Table 3 Total hazard ratio of the sampled buildings

Site	Location	HR		THR	THRI
		$PM_{2.5}$	CO_2		
A1	Ground floor	2.1	1.6	3.6	3.2
	First floor	1.8	1.0	2.8	
	Total	3.9	2.6	6.5	
A2	7th floor	0.8	2.3	3.1	3.1
	11th floor	1.8	1.4	3.2	
	Total	2.6	3.7	6.3	
A3	Ground floor	0.9	1.0	1.9	1.9
A4	Ground floor	1.3	1.9	3.2	3.2

10 Conclusion

The results of IAQ monitoring and assessment study conducted in four air-conditioned office buildings in Delhi NCR concludes that all the studied four buildings were violating the standards for CO_2 and $PM_{2.5}$ levels as prescribed by ASHRAE and NAAQS, respectively. The higher $PM_{2.5}$ in all buildings was due to higher ambient outdoor concentrations. The buildings with more proximity to the roads (building A1) and with more prominent indoor sources (building A4) were having comparatively higher $PM_{2.5}$ concentration compare to others. Poorly maintained air conditioning and ducting system (building A1) were also one of the main causes of higher indoor $PM_{2.5}$ levels. The results of CO_2 monitoring also conclude that indoor CO_2 concentrations are surrogate index of ventilation in occupied indoor spaces as the building with highest occupancy (A4) found to have highest CO_2 levels [21, 22]. Exhaust fans and proper shafts and vents in the building design (building A3) helps circulation of air and in reducing the CO_2 levels. The total hazard ratio indicator was found to be equally high in building A1 and A4. However, the impacts of poor IAQ on occupants will depend upon the sources of $PM_{2.5}$ inside the building if compare with equal THRI due to CO_2 concentration. Further research study is needed to identify the sources of $PM_{2.5}$ inside the selected indoor microenvironments.

References

1. EPA Indoor air quality (2013) http://www.eea.europa.eu/signals/signals-2013/articles. Accessed on Feb 2018
2. Fuentes-Leonarte V, Ballester F, Tenias JM (2009) Sources of indoor air pollution and respiratory health in preschool children. J Environ Public Health. 727516. https://doi.org/10.1155/2009/727516
3. Ponsoni K, Raddi M (2010) Indoor Air quality related to occupancy at an air-conditioned public building. Braz Arch Biol Technol 53(1):99–103
4. Nongkynrih B, Gupta S, Rizwan S (2018) Air pollution in Delhi: its magnitude and effects on health. Retrieved 13 Mar 2018
5. World Health Organization (2014) WHO air pollution database by country and city. World Health Organization. www.who.int/quantifying_ehimpacts. Accessed on 13th Mar 2019
6. Krewski D, Jerrett M, Burnett RT, Ma R, Hughes E, Shi Y, Turner MC, Pope CA III, Thurston G, Calle EE, Thun MJ (2009) Extended follow-up and spatial analysis of the American Cancer Society study linking particulate air pollution and mortality. HEI Research Report 140. Health Effects Institute, Boston, MA
7. Kumar P (2012) Footprints of airborne ultrafine particles on urban air quality and public health. J Civil Environ Eng 01(01)
8. Turiel I, Hollowell CD, Miksch RR, Rudy JV, Young RA (1983) The effects of reduced ventilation on indoor air quality in an office building. Atmos Environ 17(1):51–64
9. MiltonD K, Glencross PM, Walters MD (2000) Risk of sick leave associated with outdoor air supply rate, humidification, and occupant complaints. Indoor Air 10:212–221
10. Wyon DP (2004) The effects of indoor air quality on performance and productivity. Indoor Air 14(Suppl. 7):92–101. https://doi.org/10.1111/j.1600-0668.2004.00278.x

11. Jaakkola JJK, Reinikeinen LM, Heinonen OP, Majjanen A, Seppanen O (1991) Indoor air quality requirements for healthy office buildings: recommendations based on an epidemiologic study. Environ Int 17:371–378
12. Pansoni K, Stella M, Raddi G (2010) Indoor air quality related to occupancy at an air-conditioned public building, Brazilian Arch. Biol Technol 53:99–103
13. Goyal R, Khare M, Kumar P (2012) Indoor air quality: current status, missing links and future roadmap for India. Civ Environ Eng 2:4
14. Samal CG, Gupta D, Pathania R, Mohan S, Suresh R (2013) Air pollution in micro-environments: a case study of India habitat center enclosed vehicular parking, New Delhi. Indoor Built Environ 22:710–718
15. Indraganti M, Ooka R, Rijal H (2014) Thermal comfort in offices in India: behavioral adaptation and the effect of age and gender. In: 30th international PLEA conference. CEPT University, Ahmadabad, India
16. Cpcb.nic.in. (2018) Available at: http://cpcb.nic.in//upload/Downloads/AQI_Bulletin_20180309.pdf. Accessed on 3 Mar 2018
17. Datta A, Suresh R, Gupta A, Singh D, Kulshrestha P (2017) The indoor air quality of non-residential urban buildings in Delhi, India. Int J Sustain Built Environ 6(2):412–420. https://doi.org/10.1016/j.ijsbe.2017.07.005
18. ASHRAE Standard 62-2001 (2007) Ventilation for acceptable indoor air quality, American Society of Heating, Refrigeration, and Air-conditioning Engineers, Atlanta, USA
19. Christopherson RJ, Webster A (1972) Changes during eating in oxygen consumption, cardiac function and body fluids of sheep. J Physiol 221:441–457. https://doi.org/10.1113/jphysiol.1972.sp009760
20. de Gennaro G, Farella G, Marzocca A, Mazzone A, Tutino M (2013) Indoor and outdoor monitoring of volatile organic compounds in school buildings: indicators based on health risk assessment to single out critical issues. Int J Environ Res Public Health 10:6273–6291
21. Apte MG, Fisk WJ, Daisy JM (2000) Associations between indoor (CO_2) Concentrations and sick building syndrome symptoms in US Office Buildings: an analysis of the 1994-1996 BASE Study Data (LBNL 44385). Indoor Air 10:246–257
22. Ponsoni K, Maria SG (2010) Indoor air quality related to occupancy at an air-conditioned public building. Braz Arch Biol Technol 53(1):1516–8913
23. Deshpande B, Frey HC, Cao Y, Liu Z (2009) Modeling of the Penetration of Ambient $PM_{2.5}$ to Indoor Residential Microenvironment. Paper 2009-A-86-AWMA. In: Proceedings, 102nd annual conference and exhibition, 2009; Air & Waste Management Association, Detroit, Michigan
24. Allen J, MacNaughton P, Satish U, Santanam S, Vallarino J, Spengler J (2016) Associations of cognitive function scores with carbon dioxide, ventilation, and volatile organic compound exposures in office workers: a controlled exposure study of green and conventional office environments. Environ Health Perspect 124(6):805–812. https://doi.org/10.1289/ehp.1510037
25. Lappalainen S, Salonen H, Salmi K, Reijula K (2013) Indoor air particles in office buildings with suspected indoor air problems in the Helsinki area. Int J Occup Med Environ Health 26(1)
26. Seppänen OA, Fisk WJ, Mendell MJ (1999) Association of ventilation rates and CO_2 Concentrations with health and other responses in commercial and institutional buildings. Indoor Air 9:226–252
27. Satish U, Mendell MJ, Shekhar K, Hotchi T, Sullivan D, Streufert S et al (2012) Is CO_2 an indoor pollutant? Direct effects of low-to-moderate CO_2 concentrations on human decision-making performance. Environ Health Perspect 120:1671–1677

Examination of Particle Characteristics and Quantification of Emission Factors for Smoke Generated from a Popular Indian Incense Burnt in an Experimental Chamber

Anubha Goel, Radhika Mundra and Deepshikha Ola

Abstract Smoke released from the burning of incenses, a quotidian practice in India, is found to contain many toxic chemicals on particles making it a prominent source of indoor air pollution. This study analyzes the smoke and ash particles emitted from a popular incense brand in India, inside an experimental chamber. Emission factor (EF) for $PM_{3.2}$ generated from burning the incense is found to be 12.5 ± 4.2 mg/g which is higher than the EF of biomass like sugarcane, rice straw and fuelwood (range: 1.69–10.9 mg/g). This EF value is either in the same range or higher than some of the incenses from Japan, Taiwan, and Thailand. From 60 to 70% of the $PM_{3.2}$ mass collected consists of particles less than 1 μm in size. The maximum particle number count emitted from the incense exceeds 10^7 which is three orders of magnitude (i.e., 10^3) higher than the number count reported in another study from Italy (10^4). The composition of water-soluble ions and particle-bound metals in the smoke is similar to that reported for incense-based studies worldwide. This is the first study in India focusing on emission characteristics from burning incense in an experimental chamber, eliminating any external interference. Toxic elements like iron, zinc, and lead, affecting health substantially, on regular exposure are also detected. Studies have revealed that the toxicity associated with incense emissions can be higher than cigarette smoke. More comprehensive chemical analysis of the incense smoke and relevant health risk exposure is highly recommended.

Keywords Incense · Chamber study · Emission factor · Particle count

---

A. Goel · R. Mundra · D. Ola
Department of Civil Engineering, IIT Kanpur, Kanpur 208016, India

A. Goel (✉)
Center for Environmental Science and Engineering (CESE), IIT Kanpur, Kanpur 208016, India
e-mail: anubha@iitk.ac.in

A. Sharma et al. (eds.), *Indoor Environmental Quality*,
Lecture Notes in Civil Engineering 60,
https://doi.org/10.1007/978-981-15-1334-3_9

1 Introduction

Incense burning has been a part of ceremonial and ritual practices for centuries [3]. It is burnt in various forms like sticks, dhoops, cone, and joss sticks for deity worship, religious practices, and for aromatic purposes at homes [1, 6, 10].

Studies reveal that the burning of incenses can generate extensive amounts of particulate matter, chemical and gaseous pollutants [2]. The compounds detected in the smoke include fine size particulate matter, oxides of carbon and nitrogen, heavy metals, and black carbon and are known to have effects on the environment as well as the health of the people. Fine and ultrafine particles are known to be functional carriers of toxic compounds (PAHs, VOCs, and metals). These small particles penetrate more efficiently into the respiratory tract than the coarser ones and can cause biological changes inside the cells. Incense smoke is reported to be more toxic than cigarette smoke [9, 10].

The chemical composition of the incense smoke is crucial in determining the extent of the risks associated with exposure. Such studies on chemical characterization of incense smoke alone with no interferences have been lacking in India. The primary objective of this study is to quantify emissions of particles from burning of commercially popular incense in an experimental chamber. Parameters examined include particle count and mass, water-soluble ions, and elements present.

2 Methodology

A commercially popular incense was chosen according to the survey conducted in shops and information available online. Physical characteristics of the incense, i.e., average values obtained from six separate incenses from the same box, are shown in Table 1.

2.1 Experimental Setup

A stainless-steel closed chamber with a volume of 1.35 m^3 was used for collection of emissions generated from burning of incense sticks inside the chamber (Fig. 1). Loss in particle mass due to sorption on chamber walls was observed to be minimal and is assumed not to impact particle levels captured significantly. The chamber has

Table 1 Physical characteristics of the incense used for the experiment

Combustible portion		Incombustible portion	
Length (cm)	Weight (gm)	Length (cm)	Weight (gm)
14.70 ± 0.179	0.79 ± 0.031	5.48 ± 0.172	0.06 ±0.011

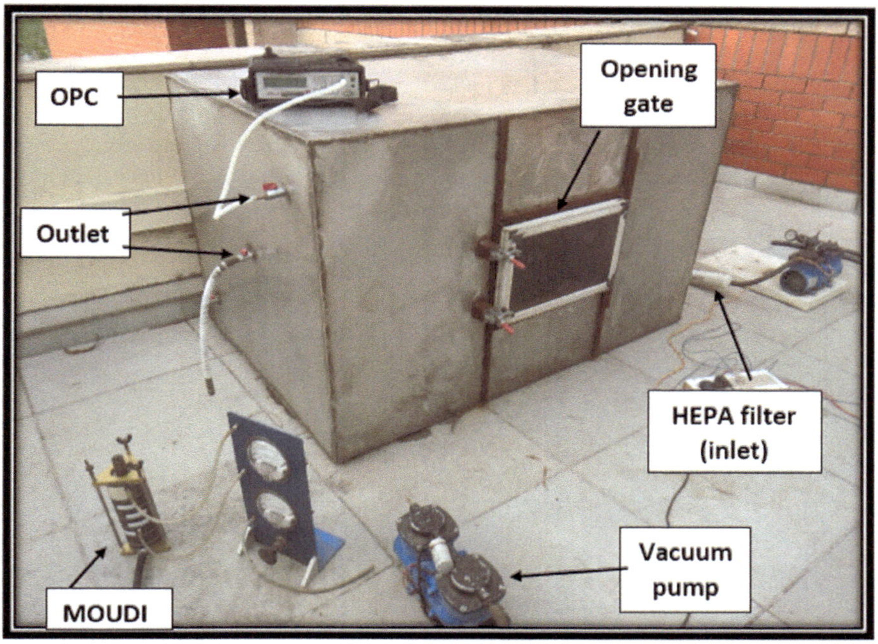

Fig. 1 Experimental Setup

three nozzles each on two opposite sides and a small gate on the third side. High-Efficiency Particulate Air (HEPA) absorber was connected on one face to clean the chamber. On the other side, Optical Particle Counter (OPC), for particle count measurement, and a cascade impactor MOUDI for the collection of particles were connected through different nozzles. A mini fan was kept inside the chamber to ensure that continuous circulation of air was maintained. The particle-free air was pushed inside the chamber before every run to purge all the particulate impurities, and particle count was monitored continuously on OPC. The process was stopped when the particle count in OPC stabilized to a minimum level of below 10^3 (the "blank value"). The burning of incense was started in the chamber after this.

2.2 *Sampling Method*

The sampling was conducted in the month of December 2017 when temperature ranged around 23 ± 2 °C. The humidity range recorded inside the chamber was 42–52%. A five-stage Cascade Impactor MOUDI and Optical Particle Counter (OPC) were used to collect the particles and to determine the number concentration of the particles emitted by the incense smoke, respectively. The size of the particles collected by MOUDI was in the range of fine particles with a diameter of 0.18–3.16 μm while the particles detected by OPC were in the range of 0.3–10 μm.

After the minimum particle count had been reached, as noted through OPC, pre-weighed incense stick was ignited and placed in the middle of the chamber on a blank paper. The blank paper was pre-weighed and used to collect the ash. The time of start and OPC reading were noted just after the closing of the door. This was denoted as "start value". MOUDI was connected to one of the outlet nozzles and was turned on approximately after 45–50 min of start time when the incense had burnt completely. With time, the particle count started decreasing to reach its "blank value". At this point, MOUDI is assumed to have collected all the particles present in the chamber emitted from burning incense. The experiment was replicated three times. Post-sampling, the incombustible incense, ash, and the filter papers were weighed to determine emission factors and for further analyses.

2.3 Sample Analysis

Filters from differnet stages of MOUDI were collectively used for the extraction of water-soluble ions (WSI) and particle-bound metals to determine the cumulative emission from 0.18–3.2 μm. The extracted samples were analyzed for water-soluble ionic concentrations of primary cations and ions (K^+, Na^+, Ca^{++}, NH_4^+, NO_3^-, SO_4^{2-}, and Cl^-) on IC (Compact IC 761 Metrohm). Elemental analysis of metals (K, Mg, Na, Co, Cr, Ni, Pb, Al, Cu, Fe, Mn, and Zn) was carried out on MP-AES (Agilent 4200 MP-AES, Agilent Technologies). The ions and elements chosen for analysis are based on the results reported in other studies.

3 Results and Discussion

3.1 Emission Factor for Particles and Ash from Incense Burning

The concentration of $PM_{3.2}$ emitted from the smoke is 9.073 ± 0.191 mg/m^3. Emission factor (EF) of particulate matter is defined as the mass of the smoke particles collected on the filter paper divided by the weight of the incense lost during burning. Similarly, EF of ash is the amount of ash obtained by burning per gram of incense. The average $PM_{3.2}$ EF obtained for the incense is 12.5 ± 4.2 mg/g, whereas for ash it is 26.4 ± 3.5 mg/g which is more than twice the EF of $PM_{3.2}$. The EF of $PM_{3.2}$ from incense is higher than EF of $PM_{2.5}$ incenses from various parts of the world including Taiwan, Thailand (11.09–23.38 mg/g), and Italy (10–26.3 mg/g) [4, 5, 9, 12]. The EF in the study is also higher than 14 out of 23 incenses experimented by Jetter et al. [3] which includes the incenses from India, USA, Thailand, and Canada.

Table 2 EFs obtained through laboratory burning of different biomasses [8]

Sources	Size	EF (g/kg or mg/g)
Forest biomass	$PM_{2.5}$	18.15 ± 14.36
Fuelwood	TSP	1.69 to 5.36
Sugarcane	$PM_{2.5}$	2.17
Rice straw (MC > 10%)	$PM_{2.5}$	4.01 to 10.9
Agricultural residue	TSP	9.31 ± 8.59
Dung cake	TSP	12.44 ± 6.13
Agricultural biomass burning in residences	TSP	8.03 ± 3.57

From a review study [8] which reports the EFs of different types of biomass burnt in laboratory conditions, it is interesting to note that the EF of the incense is much higher than many primary biomass sources like forest biomass, dung cake, and agricultural residue (Table 2). This implies that the incense burning is a crucial source of smoke emission and can be of great concern like any of the stated sources.

3.2 *Particle Mass and Count*

The size segregated analysis of the incense smoke reveals that 60–70% of the particle mass ($PM_{3.2}$) is composed of finer particles with size less than 1 μm (PM_1) (Fig. 2). This could be an underestimation since the instrument only collected particles up to 0.32 μm. The higher the percentage of finer and ultrafine particles in the smoke, the more health risk it is likely to possess.

OPC measures particle number count (PNC) in the range 0.3–20 μm. In all the test runs the particle number count exceeded the instrument's limit ($2x10^7$) within 30 minutes of the burning of the incense inside the chamber. This number exceeds the maximum PNC reported by Stabile [9] for incenses and mosquito coils by at least the magnitude of 4.

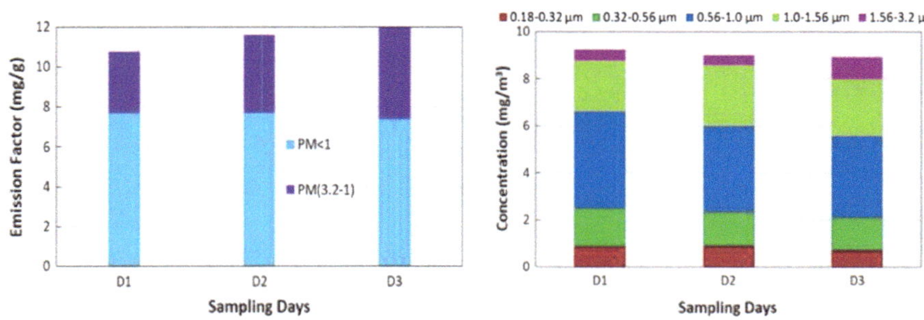

Fig. 2 Concentration of PM emitted and the % contribution of PM_1

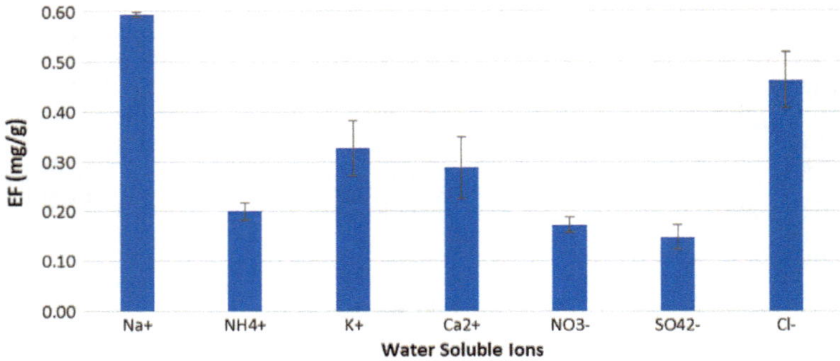

Fig. 3 Water-soluble ions in incense smoke

3.3 Water-Soluble Ions and Particle-Bound Element

The major ions detected in incense smoke emission in previous studies [7, 11] have been chosen for this study. The total EF for water-soluble ions for the incense is 2.32 $\pm$ 0.25 mg/g which is almost 20% of the total $PM_{3.2}$ mass. The water-soluble ions are predominated by sodium and chlorides with the least amount of sulfates (Fig. 3). No qualitative trend has been observed in the ionic composition of incense smoke, and it can be said that it depends on the composition of the material used for the manufacture of incenses.

3.4 Elements

Of the 13 elements analyzed, minerals K, Mg, and Na comprised almost 85% of the total mass. Ni was below the detection limit in all the cases. EFs for all the elements are presented in Fig. 4. The presence of toxic elements Pb and Cr has been reported which can cause harmful health effects even in small quantities. Zn, Fe, and Al are present in trace amounts. The EF of the total elements is 1.31 $\pm$ 0.13 mg/g (Fig. 4). The qualitative composition of the elements is similar to the emission rates from $PM_{2.5}$ reported by See and Balasubramanian [7] where the authors determined the $PM_{2.5}$ emission rates of various elements released from the incense smoke. They reported a maximum abundance of Al and Fe, after mineral elements, which is also the case in this study.

4 Conclusion

Examination of emissions from incense burning in an experimental chamber in this study reveals the significant contribution of PM_1 in $PM_{3.2}$ generated. These finer particles are likely to have more adverse effects on human health in case of extensive

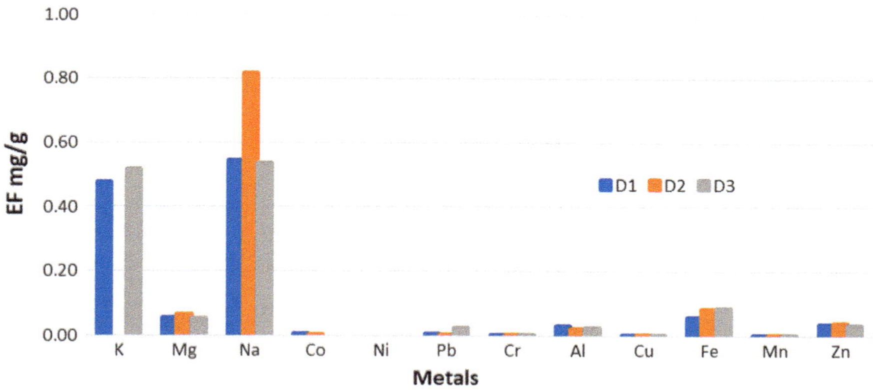

Fig. 4 EF of particle-bound elements

exposure. Emission factors of $PM_{2.5}$ are found to be higher than some prominent biomass sources bringing to attention the need for more intensive chemical characterization of the incense smoke along with other indoor pollution sources. Since no particular trend has been observed for ash and ions and differs widely among reported studies, we can say that the emissions are the source, or input material, dependent. We suggest the need to closely monitor the input materials used in manufacturing the incense and look for alternatives that are more eco-friendly on burning, i.e., they produce less smoke and acceptable levels of toxic materials.

References

1. Dewangan S, Chakrabarty R (2013) Emission of volatile organic compounds from religious and ritual activities in India. Environ Monit Assess 185(11):9279–9286
2. Fang GC, Chang CN, Chu CC, Wu YS, Fu PPC, Chang SC, Yang IL (2003) Fine ($PM_{2.5}$), coarse ($PM_{2.5-10}$), and metallic elements of suspended particulates for incense burning at Tzu Yun Yen temple in central Taiwan. Chemosphere 51(9):983–991
3. Jetter JJ, Guo Z, Mcbrian JA, Flynn MR (2002) Characterization of emissions from burning incense. Sci Total Environ 295(1–3):51–67
4. Kuo S, Tsai YI, Sopajaree K (2016) Emission characteristics of carboxylates in $PM_{2.5}$ from incense burning with the effect of light on acetate. Atmos Environ 138:125–134
5. Lee S, Wang B (2004) Characteristics of emissions of air pollutants from the burning of incense in a large environmental chamber. Atmos Environ 38:941–951
6. Liao C-M, Chen S-C, Chen J-W, Liang H-M (2006) Contributions of Chinese-style cooking and incense burning to personal exposure and residential PM concentrations in Taiwan region. Sci Total Environ 358(1):72–84
7. See SW, Balasubramanian R (2011) Characterization of fine particle emissions from incense burning. Build Environ 46(5):1074–1080
8. Simões Amaral S, Andrade de Carvalho J, Martins Costa M, Pinheiro C (2016) Particulate matter emission factors for biomass combustion. Atmosphere 7(11):141
9. Stabile L, Fuoco FC, Buonanno G (2012) Characteristics of particles and black carbon emitted by combustion of incenses, candles and anti-mosquito products. Build Environ 56:184–191

10. Suryawanshi S, Singh A, Verma R, Gupta T (2016) Identification and quantification of indoor air pollutant sources within a residential academic campus. Sci Total Environ 570:46–52
11. Wang B, Lee SC, Ho KF, Kang YM (2007) Characteristics of emissions of air pollutants from burning of incense in temples, Hong Kong. Sci Total Environ 377:52–60
12. Yang T, Lin S, Lin T, Hong W (2012) Characterization of polycyclic aromatic hydrocarbon emissions in the particulate phase from burning incenses with various atomic hydrogen/ carbon ratios. Sci Total Environ 414:335–342

Comparison of Efficiency of Active and Passive Methods of Bioaerosols' Estimation

Palak Balyan, Chirashree Ghosh, Shukla Das and B. D. Banerjee

Abstract Passive settle plate method and active impaction method are two most commonly used methods for bioaerosols' sampling and surveillance. Passive method is a relatively low-cost method for bioaerosols' surveillance. The current study intends to compare the efficiency of active and passive methods of bioaerosol sampling for temporal surveillance at different source sites. It was observed that the temporal measurements of bioaerosol by both methods were strongly correlated. Thus, both methods can be used for bioaerosol surveillance. The results of effect of putative factors by both methods were relatively comparable to a larger extent. This finding is highly significant for resource constraint setting where the use of active method is not cost effective for temporal surveillance and research purpose.

Keywords Bioaerosols · Active bioaerosol sampling · Passive settle plate method · Season and cascade impactor

1 Introduction

Microbial air surveillance has gained attention in recent years because of the improvement in knowledge regarding potential health effects associated with them. Many guidelines regarding bioaerosols' exposure levels are made by various countries and organizations [1–9].

P. Balyan (✉) · C. Ghosh
Environmental Pollution Laboratory, Department of Environmental Studies,
University of Delhi, New Delhi 110007, India
e-mail: balyan.palak@gmail.com

S. Das
Department of Microbiology, University College of Medical Sciences & GTB Hospital,
University of Delhi, New Delhi 110095, India

B. D. Banerjee
Department of Biochemistry, University College of Medical Sciences & GTB Hospital,
University of Delhi, New Delhi 110095, India

A. Sharma et al. (eds.), *Indoor Environmental Quality*,
Lecture Notes in Civil Engineering 60,
https://doi.org/10.1007/978-981-15-1334-3_10

Increasing incidences of nosocomial, gastrointestinal infections had made regular surveillance of bioaerosol in hospital setting, food processing, and pharmaceutical industries a necessary requirement. Studies have correlated nosocomial infection rate with number of airborne bacteria [10]. A proper identification and quantification of bioparticles in the air are thus necessary for chalking out effective preventive and control strategies for aerosol-borne infections.

Air sampling enables monitoring of microbial load in an environment [10]. Moreover, the microbial surveillance can also be used to check the efficiency of insulating mechanisms, disinfectant, and sterilization techniques [10, 11]. However, standards of air sampling are yet to be firmly established. In fact, international standards offer various techniques with different samplers, thus leaving ambiguity regarding the choice of technique for microbiological surveillance [12, 13].

The different methods of counting microbes which are in use are the count of colony-forming units per cubic meter of air (CFU/m^3); the count of CFU on settle plates; measurement of a chemical component of the microbial cells/m^3 of air; the count under the microscope [11].

At the moment, the most common and widely used method of quantifying airborne microbes is limited to the count of CFU [14]. The CFU count measures the microbes which are viable, can grow, and multiply. Air sampling to measure CFU is conducted in two ways: by active bioaerosol sampling or by passive bioaerosol sampling (settle plate method).

1.1 Active Bioaerosol Sampling

In active sampling, a fixed volume of air is drawn by the sampling device and passed through a collection medium. Bioaerosols thus captured are cultured and quantified as CFU/m^3 of air [10, 11]. The active method of sample collection allows sampling of aerosols independent of its size, inertia, etc., and thus able to capture the smaller particle missed by passive method due to low inertia. However, active samplers are expansive, need regular calibration and sterilization. Further, a sizable number of aerosols lose their viability during active sampling, reducing the observable bioaerosols' count.

1.2 Passive Bioaerosol Sampling: Settle Plate Method

Passive air sampling is based on gravitational force. Petri dishes having a suitable solid medium are exposed open to air for a stipulated period. Microbes fall onto the Petri plate based on their inertia. Plates were incubated at an appropriate temperature for optimum duration. Colonies grow in the plate proportional to microbial level of the air.

Settle plate is a convenient and inexpensive method and allows to obtain data from multiple places, by different operators simultaneously, thereby opening avenues for designing studies requiring large-scale bioaerosol sampling in a cheap, inexpensive, and reliable way [15].

The settle plate method is said to be biased toward selective collection of those particles large enough to be pulled by gravity or impacted by turbulence onto collecting surface. This property, in fact, is advantageous in certain situations such as operation theater, pharmaceutical, and food processing industry. Charnley [16] and French [17] favored settle plate method in comparison to active sampling in the operation theater where type and number of bacteria falling on the wound and instrument are of prime importance.

Conflicting results were obtained by various studies conducted to compare the relative efficacy of active and passive methods for bioaerosol monitoring. Thus, it is imperative to assess the effect of sampling technique on monitoring temporal variation of bioaerosol exposure and on understanding the role of various putative factors affecting bioaerosol exposure [12, 13, 18–20].

To achieve this, the current study intends to compare efficiency of active and passive methods of bioaerosol sampling for temporal surveillance and to understand the effect of putative factors on bioaerosols' levels at two different source sites.

1.3 Methodology

Two sites namely health center and central library of a university were selected for the study. Sampling of bioaerosol was conducted in the waiting hall of out-patient department (OPD) of the health center and in the main hall of central library.

1.4 Sampling Procedure

The experiments were set up so as to sample air using both the methods namely active Anderson cascade impactor and passive gravitational plate methods for a duration of 20 min. The culturable count of aerosolized bacteria and fungi was measured at center of each sampling site. The samplings were conducted weekly from May 2016 to April 2017. The samples were collected between 12 o'clock and 2 o'clock at sampling sites on consecutive days in order to minimize variation in seasonal patterns, meteorological parameters, etc., and to avoid diurnal variation. The Anderson impactor and Petri plate were placed at a distance of minimum 3 m in order to avoid the effect of turbulence created by Anderson impactor. The normal activities were allowed to be carried out during sampling so that the observed culturable count remained near to real culturable count. The colony counts of viable bacteria and fungi were read after 48 h and 72 h, respectively.

1.5 Statistical Analysis

The mean of culturable counts measured by both methods were calculated to analyze the trend of seasonal variation and variability during different sampling spell within a season, respectively. The normality and homogeneity of variance of total viable count (TVC) data were checked by conducting Shapiro-Wilk test and Levene's test. Since the data was unbalanced and had significant homogeneity of variance, the mixed model analysis of statistical package SPSS 23 was used to analyze the effect of season and site, by both active and passive methods. Since the objective of the study was to do a comparative analysis of TVC across different seasons and sites, both season and sites were entered in mixed model as fixed effects. Pearson's correlation coefficients were measured between meteorological variables and bioaerosols' levels measured by both methods. Correlations were determined to compare the efficiency of both methods in estimating the effect of putative factors on microbial counts. The Pearson correlation coefficients were calculated between the culturable counts of microbes measured by both methods in order to compare the relative efficiency and reliability of passive settle plate method relative to active impaction method.

2 Results and Discussion

The bioaerosol sampling was conducted by two methods simultaneously, the passive settle plate method and active Anderson cascade impactor method at both sites. The average counts of bacteria and fungi at each site are shown in Figs. 1 and 2.

There were more bacteria at the health center compared to the central library (Fig. 1a, b), whereas the fungal concentration at health center and library was comparable (Fig. 2a, b).

The microbial (both bacterial and fungal) counts had not shown any seasonal variation in the central library, whereas a definitive seasonal trend was observed in the health center. The minimum microbial load in air at health center was noted during pre-monsoon season followed by a rise in load during monsoon and post-monsoon seasons. Microbial load was maximum during winter at health center. The central library had not shown any conspicuous seasonal variation in both bacterial and fungal levels, and no typical pattern was observed.

The spatial and seasonal trends observed by both methods of sampling were thus comparable though the level varied as denomination unit and mechanism of sampling differed with the method.

The fixed effect of season (**$p < 0.01$ at health center and $p > 0.05$ at central library**) and source site (**$p < 0.01$**) on aerosolized bacteria count showed similar statistical significance by both methods (Table 1). The fixed effect of source site on aerosolized fungal count was statistically non-significant by both methods ($p > 0.05$) (Table 1).

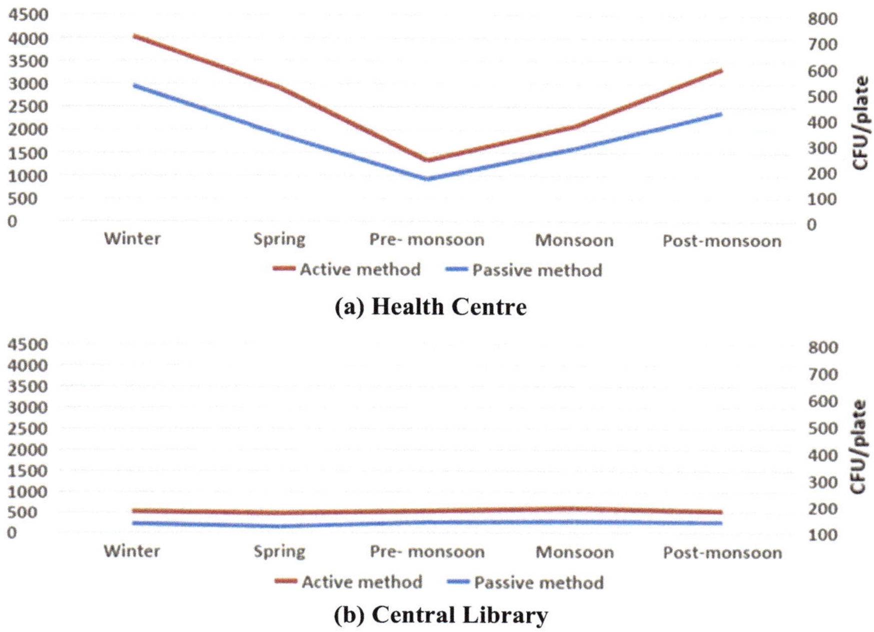

Fig. 1 Bacterial counts at health center and central library by both passive and active methods of bioaerosols' sampling

The fixed effect of season on aerosolized fungal count was significant at the health center (**$p < 0.01$**) but not at the central library (**$p > 0.05$**). Hence, microbial counts in the air samples and their seasonal variation at both sites varied with the location. However, the spatial and seasonal trends observed at each site were similar by both methods.

The effect of meteorological variables on microbial counts, calculated by both methods, was also compared, as shown in Table 2. The correlation between the microbial count and meteorological parameters was not uniform among two sites. The correlations were very weak for microbial counts at central library by both methods of sampling. Pearson's coefficient was negative between microbial count and temperature, and statistically significant strong positive correlations were observed between microbial count and relative humidity at health center.

The correlation between the bacterial and fungal counts measured by passive plate method and Anderson cascade impactor at both sites is presented in Fig. 3. The correlations were statistically significant for both bacteria and fungi at both sites. Strong correlations were noted between the bioaerosol level obtained by two methods of samplings (**ranged from $r = 0.61$ for bacteria at the central library to $r = 0.84$ for fungi at the health center**) despite different influences of temperature and relative humidity noted at different land use sites (significant correlation at health center and non-significant correlation at central library) (Fig. 3).

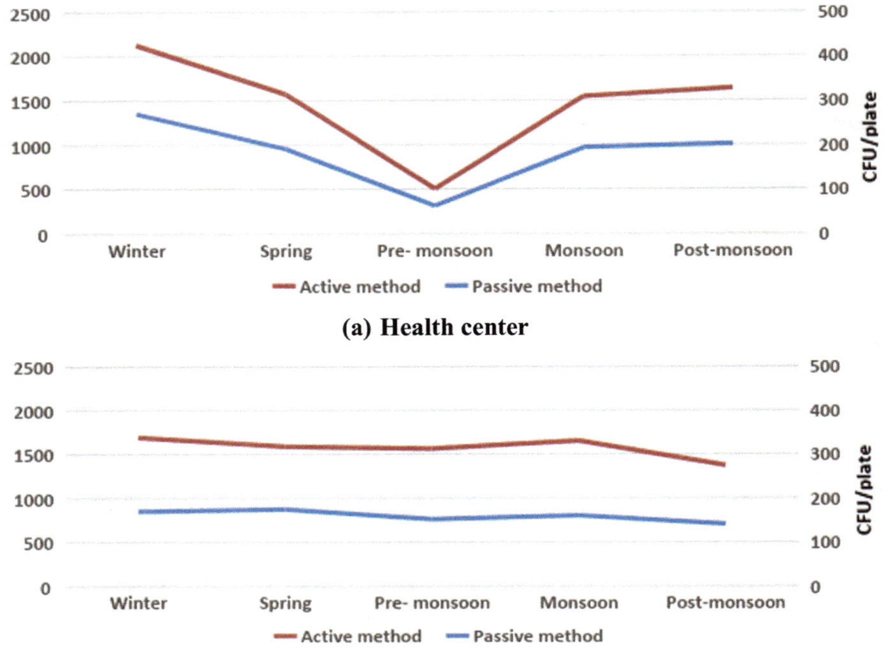

(b) Central Library

Fig. 2 Fungal counts at health center and central library by both passive and active methods of bioaerosols' sampling

Table 1 Fixed effect of season and site on microbial counts

Site	Passive method		Active method	
	Season	Site	Season	Site
Bacterial counts				
Health center	<0.01	<0.01	<0.01	lt;0.01
Central library	>0.05		>0.05	
Fungal counts				
Health center	<0.01	>0.05	<0.01	gt;0.05
Central library	>0.05		>0.05	

Significant level $p < 0.05$

Anderson cascade impactor method had captured more microbes in this study as also noted in various previous studies conducted earlier. However, this fact does not edge the use of active method as a preferred one. The quantitative comparison between the two methods is not justified in light of fact that the two methods use different denominating units and are based on different principles [21].

Table 2 Correlation matrix of bacterial and fungal counts with temperature and relative humidity

Sites	Bacteria				Fungi			
	Passive		Active		Passive		Active	
	r-value	*p*-value	*r*-value	*p*-value	*r*-value	*p*-value	*r*-value	*p*-value
Temperature								
Health Centre	−0.76	<0.01	−0.73	<0.01	−0.74	<0.01	−0.67	<0.01
Central library	0.24	0.84	0.30	0.30	0.03	0.83	0.09	0.49
Relative humidity								
Health Centre	0.32	0.01	0.30	0.02	0.53	0.01	0.48	0.04
Central library	0.19	0.16	0.04	0.76	−0.06	0.05	−0.08	0.54

Significant level $p < 0.05$

The settle plate method could be more informative in hospital wards and food processing units where settling rate of microbes is more important than their presence in the air.

In light of our findings, it can be endorsed that either method can be used for spatiotemporal monitoring of air contamination for surveillance programs or for conducting research on understanding the influence of various factors on bioaerosol levels. Further, the results of past research conducted by either of the methods can be compared and analyzed together provided that denomination units have no bearing on objectives of research work. It justifies the use of passive plate method for bioaerosol monitoring as a cheap and reliable method especially for spatial and temporal comparison. Studies by Napoli [10] and Saha [22] also reported similar correlation between bioaerosols' levels obtained by both methods. On the contrary, Sayer [20] reported active sampling to be superior to passive sampling. There is skewed distribution of studies with regard to sampling methods used in favor of active methods. It could be due to the fact that most of bioaerosol research studies are being conducted in the developed countries. There is paucity of bioaerosols' research in developing and low-income countries. In view of this fact and findings of the present study, further research for standardization of passive method of bioaerosols' monitoring is hence recommended.

3 Conclusion

The present research intended to study the relative efficiency of passive and active method of bioaerosol surveillance. The study concludes that the temporal measurement of bioaerosol by both methods was strongly correlated. Thus, both methods can

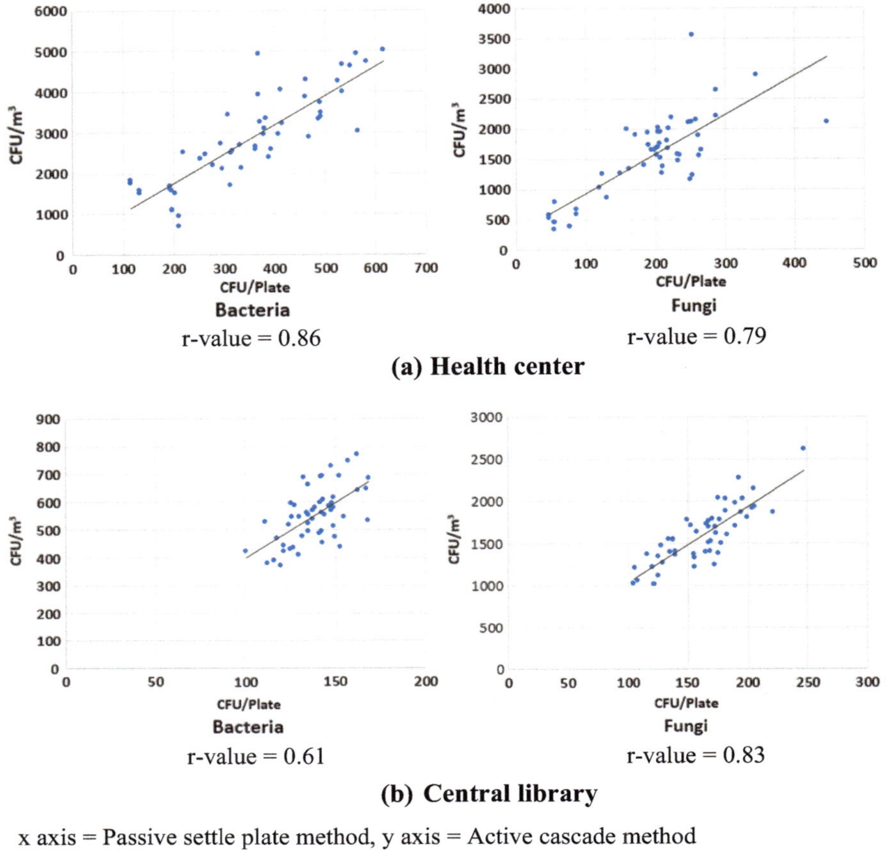

Fig. 3 Correlation matrix of microbial counts measured by active and passive methods of bioaerosols' sampling

be used interchangeably for bioaerosol surveillance. The results of effect of putative factor by both methods are relatively comparable to a large extent. This finding is highly significant for resource constraint setting where the use of active method is not cost effective for temporal surveillance and research purpose. Research for standardization of low-cost passive plate methods to develop standard guidelines is thus highly recommended.

Acknowledgements Authors are thankful to Department of Science and Technology (DST) Purse Grant and Ministry of Environment, Forest and Climate Change (MoEF), Government of India, for supporting the research work financially.

References

1. Heseltine E, Rosen J (2009) WHO guidelines for indoor air quality: dampness and mould. WHO Regional Office, Europe
2. Wanner HU, Gravesen S (1993) Biological particles in indoor environments: European collaborative action. Indoor air quality & its impact on man. Report No. 12. Commission of the European Communities, Luxembourg
3. Obbard JP, Fang LS (2003) Airborne concentrations of bacteria in a hospital environment in Singapore. Water Air Soil Pollut 144:333–341
4. Dumała SM, Dudzińska MR (2013) Microbiological indoor air quality in Polish schools. Ann Set Environ Prot 15:231–244
5. de Aquino Neto FR, de Góes Siqueira LF (2004) Guidelines for indoor air quality in offices in Brazil. Proc Healthy Build 4:549–554
6. Nunes ZG, Martins AS, Altoe AL, Nishikawa MM, Leite MO, Aguiar PF, Fracalanzza SE (2005) Indoor air microbiological evaluation of offices, hospitals, industries, and shopping centers. Memórias do Instituto Oswaldo Cruz 100:351–357
7. Bartlett KH, Lee KS, Stephens G, Black W, Brauer M, Copes R (2003) Evaluating indoor air quality: test standards for bioaerosols. Report to The Workers' Compensation Board of British Columbia Research Secretariat 99FS-64
8. Pegas PN, Evtyugina MG, Alves CA, Nunes T, Cerqueira M, Franchi M, Pio C, Almeida SM, Freitas MD (2010) Outdoor/indoor air quality in primary schools in Lisbon: a preliminary study. Quim Nova 33:1145–1149
9. Jo WK, Seo YJ (2005) Indoor and outdoor bioaerosol levels at recreation facilities, elementary schools, and homes. Chemosphere 61:1570–1579
10. Napoli C, Marcotrigiano V, Montagna MT (2012) Air sampling procedures to evaluate microbial contamination: a comparison between active and passive methods in operating theatres. BMC Public Health 12:594
11. Pasquarella C, Albertini R, Dall'Aglio P, Saccani E, Sansebastiano GE, Signorelli C (2008) Air microbial sampling: the state of the art. Igiene e sanitapubblica 64:79–120
12. Whyte W (1986) Sterility assurance and models for assessing airborne bacterial contamination. PDA J Pharma Sci Tech 40:188–197
13. Orpianesi C, Cresci A, La Rosa F, Saltalamacchia G, Tarsi R (1982) Evaluation of microbial contamination in a hospital environment. Comparison between the surface air system and the traditional method. Nuoviannalid'igiene e microbiologia 34:171–185
14. Sabharwal ER, Sharma R (2015) Estimation of microbial air contamination by settle plate method: are we within acceptable limit. Sch Acad J Biosci 3(8):703–707
15. Balyan P, Ghosh C, Das S, Banerjee BD (2017) Spatial variation of biogenic aerosols at different land use configurations in urban Delhi. Int J Appl Environ Sci 12:731–744
16. Charnley J (1972) Postoperative Infection after total hip replacement with special reference to air contamination in the operating room. ClinorthoRelat Res 87:167–187
17. French MLV, Eitzen HE, Ritter MA, Leland DS (1980) Environmental Control of microbial contamination in the operating room. In: Wound healing and wound infection. New York, pp 254–261
18. Perdelli F, Sartini M, Orlando M, Secchi V, Cristina ML (1999) Relationship between settling microbial load and suspended microbial load in operating rooms. Annali di igiene: medicinapreventiva e di comunita 12:373–380
19. Petti S, Iannazzo S, Tarsitani G (2002) Comparison between different methods to monitor the microbial level of indoor air contamination in the dental office. Annali di igiene: medicinapreventiva e di comunita 15:725–733
20. Sayer WJ, MacKnight NM, Wilson HW (1972) Hospital airborne bacteria as estimated by the Andersen sampler versus the gravity settling culture plate. Am J ClinPathol 58:558–566

21. Mandal J, Brandl H (2011) Bioaerosols in indoor environment-a review with special reference to residential and occupational locations. Open Environ BiolMonit J 4:83–96
22. Saha R, Agarawal S, Khan AM (2017) Air sampling procedures to evaluate microbial contamination: A comparison between active and passive methods at high-risk areas in a Tertiary Care Hospital of Delhi. J Patient Saf Infect Control 5:18–23

Impact of Lighting on Performance of Students in Delhi Schools

Pratima Singh, Renu Arora and Radha Goyal

Abstract Importance of lighting on human health and performance is well established. Optimal lighting conditions help in enhancing the performance and well-being of people. Whereas, glare caused by bright and direct lighting may become source of discomfort and impair visual performance. A very little research data is available correlating lighting with performance of young children in schools. Realizing the need of research in this area, the present cross-sectional study (2016–2017) investigated the impact of classroom lighting on the performance of 738 students, studying in schools in Delhi, selected through systematic sampling technique. The classroom lighting levels were represented through illuminance data recorded using lux meter 'Testo-545'. Students' performance was assessed by d2 test for speed and accuracy developed by Brickencamp and Zilmer. Questionnaire-cum-interview was used as data tool for assessing students' perception towards classroom lighting. The results conclusively revealed that classroom lighting had significant impact on the concentration and performance of the students ($P < 0.05$). Classroom illuminance between 250 and 500 lux was linked with increased concentration of the students, which translated into higher scores and increased performance. Based on the results it is recommended to increase natural light inside classrooms through strategically placed windows and doors or artificial light through appropriate lamps and luminaires. The findings further indicate the need for extensive and detailed research in the area of lighting in educational spaces.

Keywords Lighting · Illuminance · Indoor environmental quality · Glare · School · Performance · Perception

P. Singh (✉) · R. Arora
Institute of Home Economics, University of Delhi, F-4, Hauz Khas Enclave, New Delhi 110016, India
e-mail: pratima.singh@ihe.du.ac.in

R. Goyal
Indian Pollution Control Association, Hargobind Enclave, New Delhi 110092, India

A. Sharma et al. (eds.), *Indoor Environmental Quality*, Lecture Notes in Civil Engineering 60, https://doi.org/10.1007/978-981-15-1334-3_11

1 Introduction

Good quality lighting is critically important to carry out tasks efficiently and safely in work and educational environment. Significance of lighting is widely researched and claimed to be vital for occupants' performance. Classroom lighting does not only help students to see instruction-board clearly but also assist them to comfortably read and write during classroom learning activities [1]. Also, lighting helps to create visually stimulating environment which enhances the appearance of the indoor space [2]. Research studies have shown influence of lighting on biological health of humans such as circadian rhythm, heart rate, and blood pressure. Lighting may stimulate certain psychological behaviours like mood swings. Besides emotional and biological effects, studies have indicated positive effects of lighting conditions on speed and accuracy of completion of tasks and productivity of the occupants [3, 4]. However, most of these empirical studies have focused on offices, retail, and other work environments [4]. Research data on the effect of lighting in educational spaces are sparse even today. Understanding the relationship of lighting and environment can help to improve indoor environmental quality for better performance of students.

1.1 Impact of Lighting on Human Health

Interest of researchers on the effects of lighting on physical and psychological health of human being is rapidly advancing. The positive health effects of natural light were even acknowledged by Egyptians, Romans, and the ancient Greeks, who worshipped the Sun Gods. Even in India during the Mughal era, environment-friendly techniques of using *Jaali-Numa* (perforated) walls or windows were used to increase the entry of daylight and control glare inside the buildings [5]. The significance of lighting goes beyond its functional aspect.

1.1.1 Effect on Physical Health

Lighting must be designed in a human-centric way, keeping in mind the tasks to be performed and comfort of all the occupants. Many research studies have claimed that insufficient illuminance, flickering of lights, and absence of lighting controls in buildings can lead to health problems like eye strain, headache, increased body temperature, etc. [6, 7]. On the other hand, excessive light and luminance contrasts can cause glare which can cause eye strain, blurred, and weak vision.

Effect on Cognitive Health and Circadian Rhythm

Light is known to control circadian rhythm in humans, by regulating stress hormone cortisol and sleep hormone melatonin. These hormones control an organism's internal clocks [8] (Fig. 1). Researches have indicated the effect of inadequate daylight on

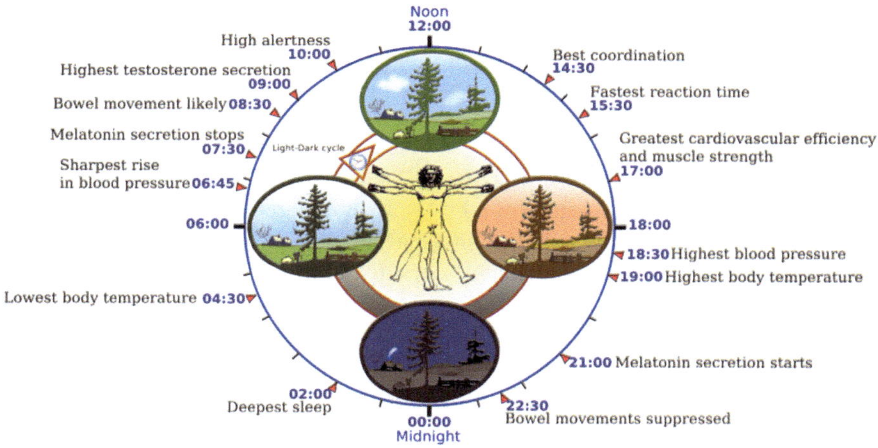

Fig. 1 Typical circadian pattern in humans [10]

reduced circadian cycle, as a result of which melatonin is secreted at wrong times causing chronic fatigue, depression, and other illnesses. Children need to receive adequate light in the morning hours and avoid light with similar wavelength in the evening to maintain a regular sleep cycle pattern. As majority of their daytime is spent in schools, the windows of the classrooms must be designed to allow sufficient daylight, as natural light helps in synchronizing internal body clock of human beings. Daylight controls circadian rhythmicity which regulates many cognitive processes in humans such as attention, concentration, and memory [9].

Psychological Effects of Lighting
Other than physical health effects, the psychological impact of lighting is more profound. Appropriate lighting and colours may uplift mood and make people feel joyful and happy [11]. On the other hand, poorly lit spaces may cause feeling of dullness, irritation, lethargy, and depression. The seasonal depression usually profound amongst people living in northern latitudes, also referred to as seasonal affective disorder (SAD), which is caused by lack of daylight and is an obvious evidence which proves the psychological effects of natural light on humans [12]. Furthermore, many research studies indicate that presence of daylight enhances mental performance and decreases aggressive behaviour [13].

1.2 Lighting and Performance

Provision of adequate lighting in learning places is important because of the direct relationship of good lighting with performance of the students [8, 14]. The purpose of lighting in educational spaces is to create visually stimulating environment in which the act of learning can be done in the best way with minimal discomfort. A study by Heschong Mahone Group indicated positive correlation between daylight and academic performance of the students [15]. It is well recognized that good quality lighting increases the comfort of students and that comfort often translates into higher scores, increased performance [16, 17], and concentration [4].

However, more lighting does not always have positives effects on occupants. Excessive light, flickering of bulbs or tube-lights, and glare produced by the fluorescent lighting and interactive boards can become the cause of discomfort, headaches, and impaired vision, and may influence cognitive performance of the students in schools [7]. A Research conducted in Brazilian classrooms provided an upper limit to classroom lighting, above which the lighting had negative effects [18]. Along with the appropriate lighting levels, uniformity of lighting throughout the classroom is also important to create pleasant and comfortable work environment.

Significance of natural or daylighting is acknowledged more than artificial lighting [19, 20]. However, when daylight is not sufficient to provide comfortable visual environment, it must be supplemented with artificial light [21].

1.3 Lighting Recommendations and Guidelines for Educational Spaces

Lighting and illuminance are the most important physical parameters of indoor environmental quality of any school or classroom. Carefully designed artificial light and provision of adequate natural light are prerequisites of any intellectual learning space. As cited by Pulay (2010), Benya defined a well-lit classroom as the one which includes controlling glare, having balanced brightness and higher reflectance [22]. Students engage in varied activities in classrooms like reading on desks, learning from projectors and blackboards, and thus, they often have to shift their gaze up and down. Therefore, all-around optimal lighting on walls as well as on desks is very critical and important [22, 23]. While designing the learning spaces focus on appropriate lighting, occupants' comfort and energy usage must be kept in mind for providing optimal visual environment.

Many research studies suggest that for appropriate lighting and visual comfort, the illuminance of school classrooms must be 300 lx or more at any point on the work surface. However, where lighting of a space is achieved by a combination of daylight and artificial light, the minimum illuminance of 350 lx is recommended in ATL-the Education School Premises Regulations, 1981 [24]. Similarly, National Building Code (NBC-2005) of India has recommended a range of illuminance (200 lx-300 lx-500 lx) for all type of classroom activities [25]. However, NBC does not provide recommendations for specific classroom-related tasks. The European norm EN 12464-1 provides guidelines for maintaining task-based illuminations inside classrooms [26]. Table 1 illustrates the summary of lighting and illuminance recommendations and guidelines for schools and classrooms given by various organizations.

Table 1 Illuminance recommendations and guidelines for schools and classrooms

Standard	Recommended illuminance (in lx)	Task	Source
National Building Code, 2005, India	200-300-500	All classroom tasks	[25]
European norm EN 12464-1	500 300 300	Blackboard tasks Writing, reading, drawing Looking at projector screen	[26]
ATL-The School Education Premises Regulations, 1981	350	All classroom tasks	[24]
CIBSE-2006	300 500	General teaching spaces Seminar rooms, teaching spaces, and lecture halls	[7, 27]

Though it is evident from the reviewed literature that lighting has an important role to play in indoor environment, availability of evidences on the extent to which lighting affects school performance of young children is still sparse and more attention of researchers needs to be drawn in this area. Hence, the present research was designed to study the impact of classroom lighting on the performance of the students in schools located in Delhi.

2 Methodology

This cross-sectional research study tried to identify the impact of classroom lighting on students' perceptions and performance. The data was collected from four private schools located in Delhi (Fig. 2). The research encompassed of the specific objectives of studying the design features of the classrooms of the selected schools; monitoring existing lighting/illuminance in the classrooms of the selected schools; and assessing students' perceptions and satisfaction regarding their classroom lighting conditions. Students' concentration and performance (CP) was also assessed to find out the impact of classroom lighting on their performance scores (CP scores).

The information regarding students' perception was assessed through structured questionnaire-cum-interview tool administered on 738 students (384 students in non-winter season and 354 students in winter season) selected through systematic sampling technique. Further, students' performance and concentration were assessed by using a standardized d2 test, which was created by Brickencamp and Zilmer in 1998. The test is also known as cancellation task and takes approximately 8–10 min to complete. The test consisted of 14 rows of 47 randomly mixed letters: p or d. In each line, the letters d and p were printed with each up to two marks above or underneath the letters. All the d's with two marks altogether were to be crossed out within

Fig. 2 Location of selected schools A, B, C, and D on map of Delhi [28]

Fig. 3 Illuminance monitoring in classroom of school B using lux meter 'Testo-545'

20 s per line. The attention score of the students was calculated from the number of d's crossed out correctly minus the number of letters crossed out incorrectly (error of commission), which is also termed as concentration performance (CP). It is a measure of coordination of speed and accuracy of performance.

The lighting levels were represented through illuminance data collected from three classrooms of each school in both non-winter and winter seasons during the year 2016–2017. The illuminance monitoring was done thrice a day, i.e., at 9:00 am,

Table 2 General observations of the selected schools

School	Classroom	Direction of windows	Area (m^2)	Window-to-floor area ratio (WFR)
A	A1	166° North-East	37.51	01:24
	A2	9° North	42.31	01:20
	A3	96° North	37.76	01:21
B	B1	214° South-West	68.89	01:12
	B2	109° South	47.19	01:06
	B3	185° South	54.72	01:06
C	C1	185° South	28.07	01:20
	C2	343° North	35.61	01:15
	C3	190° South	29.75	01:25
D	D1	204° South-West	41.37	01:26
	D2	223° South-West	41.37	01:26
	D3	258° West	41.37	01:26

at 12:00 noon and at 2:00 pm in each selected classroom using lux meter 'Testo-545' (Fig. 3). The instrument works on the principle of silicon photodiode and has measurement range from 0 to 100,000 lx with the accuracy of ±5%.

3 Results and Discussion

3.1 General Classroom Observations

Structured checklist was used for studying design features of the classrooms of selected schools. The results highlighted that the area of the classrooms ranged from 28.07 to 68.89 m^2. It was also found that window-to-floor area ratios (WFRs) of classrooms of schools A, C, and D were well within recommended range (more than 1:10), except school B (Table 2). School B classrooms had very narrow windows limiting the entry of daylight. The glass used on the windows was also glazed, and hence entry of sunlight was restricted in school B classrooms.

3.2 Comparison of School-Wise Lighting Data with Recommended Limits

It was observed that the highest mean illuminance in non-winter season was in classroom D1 (991.1 lx) and in winter season, it was in classroom A3 (1145.3 lx). The lowest mean illuminance in both the seasons was recorded in classrooms of school B,

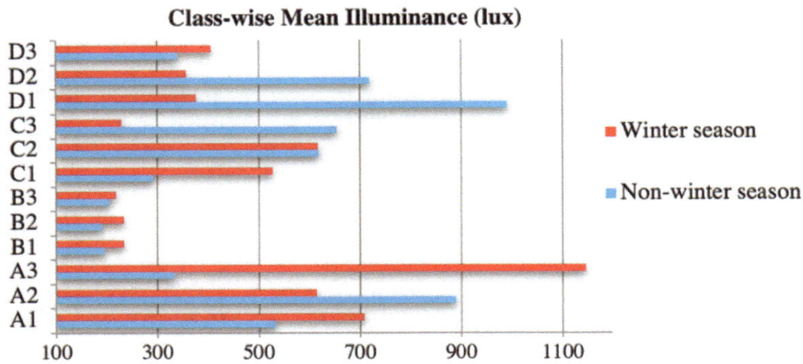

Fig. 4 Class-wise mean illuminance in non-winter season and winter season

i.e., classroom B2 had 190.2 lx in non-winter season and classroom B3 had 218.1 lx in winter season (Fig. 4). Overall, it was observed that in winter season, approximately 58% classrooms had lighting levels within the prescribed range (200–500 lx) as suggested by NBC-2005. However, in non-winter season, only 33% classrooms had lighting within recommended range. Almost 50% of the classrooms were over illuminated.

NBC-2005 has recommended lighting/illuminance levels between 200 and 500 lx to carry out all type of classroom activities comfortably [21]. Illuminance levels inside the schools were analyzed in detail to assess the existing lighting conditions inside the school classrooms and whether they fall within the recommended range or not. In non-winter season, the mean values of illuminance inside schools were 584.8 lx (school A), 197.0 lx (school B), 521.7 lx (school C), and 703.1 lx (school D); whereas, in winter season the observed mean values of illuminance were 661.3 lx (school A), 228.0 lx (school B), 458.2 lx (school C), and 380.0 lx (school D).

On comparing lighting of schools in winter season with the recommended limits suggested in NBC-2005 (lower limit: 200 lx, upper limit: 500 lx), it was found that mean illuminance of schools B, C, and D was within the recommended range and school A was over illuminated. On contrary to winter season, none of the schools met the recommended range in non-winter season. Lighting in schools A, C, and D were exceeding the upper limit and school B was poorly illuminated having lighting less than minimum prescribed limit (Fig. 5).

Due to smaller WFR and use of glazed glass on windows, the entry of natural light was limited in school B. Therefore, in both the seasons, school B had lowest mean illuminance in comparison to other schools. The results pointed out that placement and size of the windows in relation to room size or WFR significantly affects the total light available inside the classroom. The room with the lowest WFR also had lowest lighting levels. Hence, if windows are not able to provide enough daylight, there must be provision of enough artificial light in the classroom. Also, the use of window curtains can prove to be beneficial to avoid over illuminance and problem of glare in the rooms.

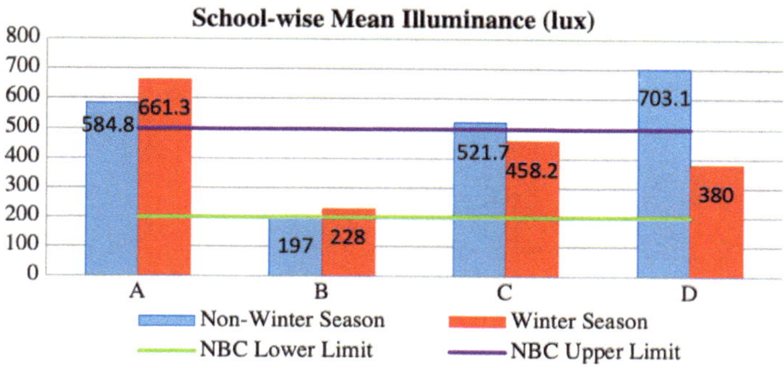

Fig. 5 School-wise mean illuminance in non-winter season and winter season

3.3 *Perception of Students Towards Their Classroom Lighting*

Researches have indicated that appropriate lighting quality can increase productivity and performance, and decrease eye strain and fatigue amongst occupants [29]. Adequate lighting is an important aspect of classroom environment as it enhances the ability of students to clearly read, write, and carry out other activities comfortably. In the present research, an attempt was also made to find out perception of students towards general lighting conditions inside their classrooms.

Analysis of the results indicated that in winter as well as non-winter season, majority of the students from schools A, C, and D felt presence of sufficient natural light inside their classrooms; whereas, in case of school B, a great percentage of students reported insufficient natural lighting inside their classrooms (32.90% in non-winter season and 18.50% in winter season) (Figs. 6 and 7). These findings were in line with the lighting (illuminance) data, wherein, the school B was also found to have lowest mean illuminance (Fig. 5) and lowest WFR (Table 2). The reason for insufficient natural lighting inside school B classrooms was placement and size of the windows in relation to the room size.

Fig. 6 Perception of students towards natural lighting inside classroom in non-winter season

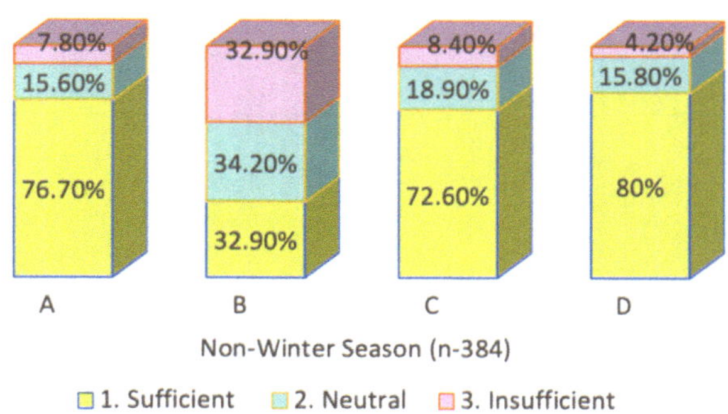

Fig. 7 Perception of students towards natural lighting inside classroom in winter season

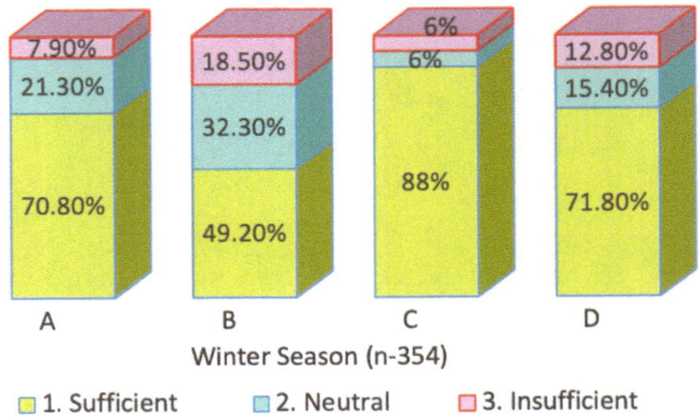

It was also found that students faced difficulty in reading or copying from the blackboard/whiteboard due to the glare caused by glossy surface or excessive contrast between the dark areas and bright areas in the direction of viewing. In extreme cases, glare can also cause eye fatigue, headache, etc., and can impair visual performance by reducing task visibility. The analysis of the results indicated that in non-winter season, a significant number of students from school B (53.2%) and from school D (41.7%) complained of the problem of glare on the blackboard; whereas, only 21.1% students from school A and 24.2% from school C complained about problem of glare. During the winter season, almost 70.8% students from school B, 55.6% from school D and 42.2% from school C complained of glare on the instruction-board (Table 3). School-wise statistically significant difference was found in the perception of students towards experience of glare inside their classroom in both the seasons ($P < 0.05$).

The reason for higher percentage of students facing the problem of glare in school B was the usage of whiteboard for instructions, whereas, in schools C and D, the reason was reflectance of sunlight on the blackboard. A very few students faced the

Table 3 Students perception towards experience of glare inside classroom

Season	Non-winter season (*n*-384)				Winter season (*n*-354)			
School	A	B	C	D	A	B	C	D
	n-90	*n*-79	*n*-95	*n*-120	*n*-89	*n*-65	*n*-83	*n*-117
Do you experience glare on the blackboard/chalkboard/whiteboard?								
No (%)	46.7	29.1	56.8	28.3	51.7	16.9	42.2	29.9
Don't know (%)	32.2	17.7	18.9	30.0	29.2	12.3	15.7	14.5
Yes (%)	21.1	53.2	24.2	41.7	19.1	70.8	42.2	55.6
Total (%)	100.0	100.0	100.0	100.0	100.0	100.0	100.0	100.0
χ^2	37.425							48.398
p-value	0.000*							0.000*

* *p*-value significant at 0.05

problem of glare in school A as this school heavily relied on projector-based teaching. School A had window screens, and whenever projector was used window screens were lowered to avoid glare. Another important finding of the study was that most of the students from the schools where classrooms faced west, south or south-west direction [i.e., school B and school D] experienced glare on the board. North-facing windows provide good quality daylight and general teaching spaces that face north receive even and consistent light in all the seasons.

3.4 *Relationship of Lighting with Performance of Students*

Results highlighted that the mean performance score (CP score) of the students of school A was 207.03, school B was 183.93, school C was 189.74, and school D was 189.06 (Fig. 8). It was also observed that schools which had higher lighting (illuminance) also had higher CP scores. The results indicated that the students' performance scores had positive correlation with classroom lighting/illuminance (*p*-value < 0.05). These results were in parity with the reviewed research studies which had also indicated a positive correlation between lighting and students' performance [4, 11].

Further, binary logistic regression was performed between the performance score (CP scores) and lighting (illuminance) at 0.05 level of significance to assess impact

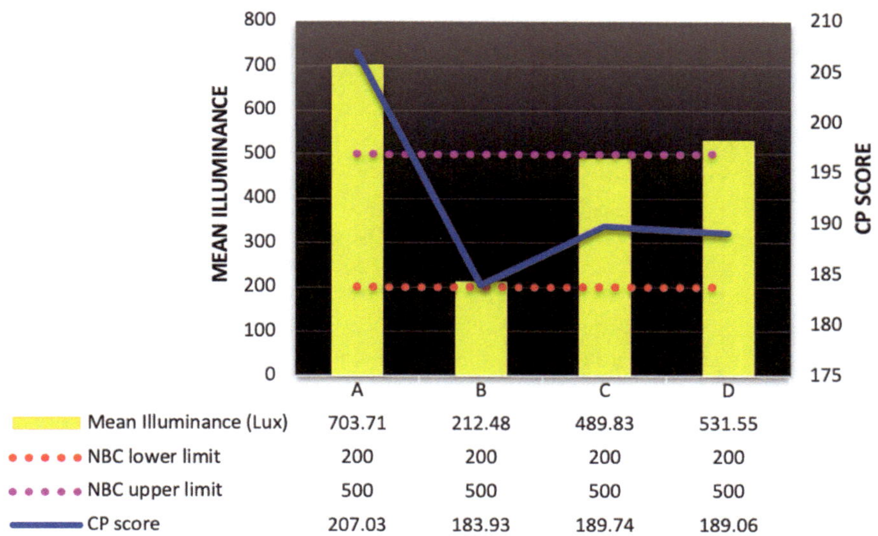

	A	B	C	D
Mean Illuminance (Lux)	703.71	212.48	489.83	531.55
NBC lower limit	200	200	200	200
NBC upper limit	500	500	500	500
CP score	207.03	183.93	189.74	189.06

Fig. 8 School-wise mean illuminance in non-winter season and winter season

Table 4 Logistic regression analysis for association of illuminance with CP scores of students

Variables	df	Sig.	Exp (B)	95% Confidence interval	
				Lower	Upper
Illuminance (0)	2	0.002*	–	–	–
Illuminance (1)	1	0.005*	2.273	1.278	4.042
Illuminance (2)	1	0.003*	1.973	1.261	3.085

df degree of freedom for Wald chi-square test, *Exp(B)* exponentiation of the B coefficient/odds ratio for predictors
*p-value significant at 0.05

of classroom lighting on performance of the students. The results indicated that classroom lighting had a high significance in affecting the outcome variable, i.e., performance score (CP score) of the students (Table 4). Classroom lighting/illuminance between 250 and 500 lx was linked with increased concentration of the students, which translated into higher scores and increased performance. The analysis also indicated that the performance of the students increased with the increase in illuminance till a certain comfort limit after which it started declining.

4 Conclusion

The study showed that strong correlation exists between classrooms lighting and performance of the students. Classroom lighting between 250 and 500 lx was linked with increased concentration of students, which translated into higher scores and improved performance [13]. Optimal lighting in schools can create favourable viewing conditions for students to read and copy from the instruction-board. Schools are advised to make effective use of window blinds or curtains in the classrooms to avoid glare caused by excessive daylight or highly reflective work surfaces which may lead to eye strain and fatigue amongst students. As classroom environment plays an important role in the overall performance of the students, and hence they need to be planned and designed according to the standards and comfort needs of the students. The findings further suggest the need for extensive and detailed research in the area of lighting in educational spaces and student learning.

Acknowledgements We acknowledge grant received from Research Council, University of Delhi for funding monitoring instrument under Research and Development Project 2013, Reference no: DRCH/R&D/2013-14/4155.

References

1. Cristian AI, Jennifer Carolina GZ, Helenice MS (2017) Natural and artificial lighting analysis in a classroom of technical drawing: measurements and HDR images use. In: Creative construction conference 2017, Procedia engineering, vol 196, pp 964–971. Elsevier, ScienceDirect
2. Future Proofing Schools, Faculty of Architecture, Building, and Planning, The University of Melbourne, (2011). https://research.unimelb.edu.au/__data/assets/pdf_file/0005/2435027/Future-Proofing-Schools-Phase-3-Research-Reflections.pdf, last accessed on 2017/01/1
3. Kyungah C, Hyeon-Jeong S (2016) Dynamic lighting system for the learning environment: performance of elementary students. Opt Express 24:A907–A916
4. Sleegers PJC, Moolenaar NM, Galetzka M, Dijk JV, Pruyn A, Sarroukh BE, Zande BVD (2012) Lighting affects students' performance positively: findings from three Dutch studies. Light Res Technol 45(2):159–175
5. Singh P (2015) Buildings for life: creating the future, conserving the past. J Civ Eng Environ Technol (JCEET) 2(3):211–213. Krishi Sanskriti Publications
6. Sahlberg B, Smedje G, Norbäck D (2002) Sick building syndrome (SBS) among school employees in the county of Uppsala, Sweden. Proc Indoor Air 2002:494–499
7. Winterbottom M, Wilkins A (2009) Lighting and discomfort in the classroom. J Environ Psychol 29:63–75
8. van Bommel IWJM, van den Beld IGJ (2003) Lighting for work: visual and biological effects. Philips Lighting, The Netherlands
9. Stevens RG, Rea MS (2011) Light in the built environment: potential role of circadian disruption in endocrine disruption and breast cancer. Cancer Causes Control 12:279–287
10. Biological clock human, https://commons.wikimedia.org/wiki/File:Biological_clock_human.svg, last accessed on 2019/04/14
11. Samani SA (2012) The impact of indoor lighting on students' learning performance in learning environments: a knowledge internalization perspective. Int J Bus Soc Sci 3(24):127–136
12. Boubekri M (2008) Daylighting, architecture and health. Architectural Press, NY
13. Edwards L, Torcellini PA, N.R.E. (2002) A literature review of the effects of natural light on building occupants. National Renewable Energy Laboratory
14. Jago E, Tanner K (1999) Influence of the school facility on student achievement. http://www.coe.uga.edu/sdpl/researchabstracts/visual.html, last accessed on 2011/03/3
15. Heschong Mahone Group (1999) Day lighting in schools: An investigation into the relationship between daylighting and human performance. http://www.pge.com/pec/daylight/valid.html, last accessed on 2014/12/29
16. Cash CS (1993) Building condition and student achievement and behaviour. Abstracts international, http://cnx.org/content/m23100/latest/, last accessed on 2014/09/19
17. Rodgers PA (1998) Case study: an elementary school that saves energy and is visually comfortable. Archit Record 186(8):159–160
18. Kruger EL, Zannin PHT (2004) Acoustic, thermal and luminous comfort in classrooms. Build Environ 39:1055–1063
19. Cajochen C (2007) Alerting effects of light. Sleep Med Rev 11:453–464
20. Heschong L, Wright RL, Okura S (2002) Daylight impacts on human performance in school. J Illum Eng Soc, 101–114
21. Barnitt H (2003) Lighting for the future. Build Serv J (The magazine for the CIBSE) 25(1):38–39
22. Pulay AS (2010) Awareness of Daylighting on Student learning in an educational facility. University of Nebraska—Lincoln, Lincoln
23. John M, Timothy EH (2005) Illuminating the classroom environment. Sch Plan Manag 44(2):34
24. ATL—The Education Union (1981) Standards for education premises. http://www.atl.org.uk/health-and-safety/work-environment/standards-education-premises.asp
25. Manuals GRIHA (2011) Manual I: Introduction to national rating system. Energy Resour Inst 3:55–123

26. Dahlan AS, Eissa MA (2015) The impact of day lighting in classrooms on students' performance. Int J Soft Comput Eng (IJSCE) 4(6):7–9
27. CIBSE Student Members' Data Book (2006) CIBSE Engineering and Sustainable Built Environment, pp 1–2. http://tomrobbo.com/dbsinfo/STUFF/CIBSE%20Student%20Data%20Sheet.pdf
28. https://en.wikipedia.org/wiki/Narela#/media/File:Delhimap.jpg, last accessed on 2019/04/14
29. One workplace, Seeing the Difference (1999) The importance of Quality Lighting in the Workplace. Workplace Issues, http://www.pdf-finder.com/the-importance-of-qualitylighting-in-the-workplace.html, last accessed on 2014/ 07/04

Exposure to Particulate Matter in Classrooms and Laboratories of a University Building

Abinaya Sekar, Prem Mohan, George K. Varghese and M. K. Ravi Varma

Abstract In India, around 1.3 million deaths occur every year due to indoor air pollution. The most common indoor pollutants include asbestos, biological pollutants, carbon monoxide, formaldehyde, lead, nitrogen dioxide, pesticides, radon, indoor particulate matter (IPM), second-hand smoke and volatile organic compounds. In this study, the concentrations of IPM_{10}, $IPM_{2.5}$ and IPM_1 were as measured using Laser Aerosol Spectrometer (Grimm MiniLAS 11-R) in the major laboratories and classrooms of the Department of Civil Engineering at National Institute of Technology, Calicut, India. The sampling was carried out during working days and found that the highest level of particulate matter IPM_{10} and $IPM_{2.5}$ were found in the concrete laboratory, where the major source for dust particles could be cement. The highest level of IPM_1 was found in the dumping yard within the structural engineering block. The IPM_{10} and $IPM_{2.5}$ levels in almost all the laboratories exceeded the permissible value prescribed by WHO. The faculty and research scholars' cabin within the laboratory space is highly prone to particulate matter pollution.

Keywords Indoor particulate matter · Classrooms · Laboratories · Laser Aerosol Spectrometer

1 Introduction

Studies conducted over the past two decades reveal that there is a very high correlation between health-related issues and the level of pollution in the outdoor and indoor environment [1–5]. Some studies have shown that in spite of the absence of sources

A. Sekar · P. Mohan · G. K. Varghese (✉)
Department of Civil Engineering, National Institute of Technology Calicut, Kozhikode, Kerala, India
e-mail: gkv@nitc.ac.in

M. K. R. Varma
Department of Physics, National Institute of Technology Calicut, Kozhikode, Kerala, India

A. Sharma et al. (eds.), *Indoor Environmental Quality*, Lecture Notes in Civil Engineering 60,
https://doi.org/10.1007/978-981-15-1334-3_12

like smoking and cooking, particulate matter pollution levels in laboratories and classrooms were higher when compared to that in residential and other commercial buildings [6, 7].

In India, particulate matter is one of the major pollutants, and the level exceeds the permissible limit in many of the major cities [8]. Indoor air quality is often influenced by outdoor air in the case of naturally ventilated buildings. Monitoring of indoor particulate matter 2.5 ($IPM_{2.5}$), indoor particulate matter 10 (IPM_{10}) and indoor particulate matter 1 (IPM_1) is very important in micro-environments like classrooms and laboratories due to their impact on health globally [9].

Maintaining good environmental quality in classrooms is part of providing a healthy and comfortable learning atmosphere for the students and faculty which in turn will influence their health, productivity, performance and comfort in a variety of ways [10]. Exposure to indoor pollutants is one of the major health issues in many fast-developing countries [11]. In India alone, around 1.3 million deaths are caused every year due to indoor air pollution. The health effects caused due to indoor air quality are usually called as building-related issues or sick building syndrome. Sick building syndrome is mainly due to poor ventilation and the high density of population in a particular area [12]. Among IPM, respirable fraction causes pneumoconiosis and silicosis, whereas larger size fraction causes bronchitis and obstructive changes in pulmonary function. Sampling and analyzing the PM_{10} and $PM_{2.5}$ concentrations give crude exposure rates but will not give accurate results to correlate to adverse health impacts associated with it. For this, a detailed analysis of the size-wise concentration of particulates can be more useful [13].

In this study, the focus was mainly on measuring IPM_{10}, $IPM_{2.5}$ and IPM_1 in laboratories and classrooms of the Civil Engineering Department (CED), National Institute of Technology, Calicut (NITC). The Department has a number of specialized laboratories which include Geotechnical Engineering, Environmental Engineering, Transportation Engineering, Coastal Engineering, Construction Engineering, Structural Engineering, Hydraulics and Water Resource Engineering, Geoinformatics and Engineering Geosciences. In many of these laboratories, the work involves the use of cement, sand, soil, chemicals and other construction materials which can contribute particulate matter concentration in the indoor environment. Since research scholars, faculties and most of the postgraduate students spend around 8–10 h every day in their laboratory workspace and cabins inside the laboratory space, it is important and necessary to frequently monitor the pollution levels and take necessary steps to control pollution in these indoor environments. In addition, studies similar to the present study will help to develop IAQ standards in India.

2 Methodology

The general outline of the study is given in the form of a flowchart in Fig. 1, and detailed description of the entire methodology is given in subsequent sections.

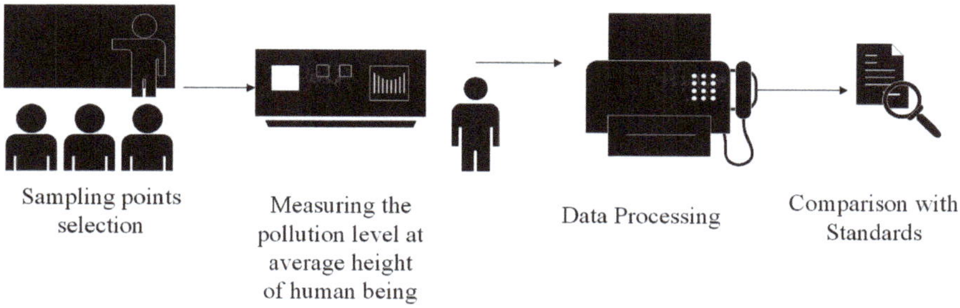

Fig. 1 General methodology

2.1 Sampling Location

Kozhikode is the second largest urban agglomeration in Kerala, India, with a population of around 2 million as per 2011 census. The city is located 250 km west of Bangalore, 235 km south of Mangalore and 525 km northeast of Chennai. NITC is located in the foothills of western ghats, 22 km northeast to the city of Kozhikode (11°19′19.00″ N and 75°56′00.8″ E). In the study, sampling was done in classrooms and laboratories of Civil Engineering Department during the month of April 2018. The Department has 582 undergraduate students, 170 postgraduate students, 53 full-time research scholars, 41 faculty members and 20 non-teaching staff. It is worth mentioning that the campus is cleaner and greener when compared with other parts of the city. The blue blocks in the map (Fig. 2) indicate the sampling blocks, and red points indicate the location where outdoor samples were collected.

Fig. 2 Monitoring site location. *Source* Wikipedia and Google map

Table 1 Instrument details [14]

Parameter	Details/value
Instrument name Model Make	Mini laser aerosol spectrometer 11-R GRIMM
Size channels	31 channels from 0.25 to 32 μm
Sample flow rate	1.2 l/min ±5%
Rinse flow rate	0.3 l/min
Reproducibility	±3% over the total measuring range
Laser wavelength	660 nm
Count range	1–2,000,000 particles/l
Particle mass	0.1 $\mu g/m^3$–100 mg/m^3

2.2 Sampling Design and Instrumentation

The sampling was carried out in 30 different locations within and surrounding the Department to measure both ambient and indoor concentrations. The instrument was kept 1.5 m above the ground surface at the breathing zone of an average human being. The monitoring was done in each location, and the mean value was tabulated. Pollution level was monitored in such a way that all the sampling points could be covered in a single day and that would help us in providing a better comparison of results. The technical information about the instrument used for the sampling is given below (Table 1).

2.3 Data Processing and Comparison with Standards

Data processing was carried out using the software Version 1.178, LabView® for further comparison with air quality standards. Since India does not have its own indoor air quality standards, the obtained results were compared with the National Ambient Air quality standard of India (NAAQS) and WHO air quality standards. In addition to that, the results were compared with European Union Standard EN 481, 1993 for inhalable fraction, thoracic fraction and respirable fraction values (Table 2).

3 Results and Discussions

The IPM_{10} concentrations exceeded Indian National Ambient Air Quality Standards (100 $\mu g/m^3$, 24-h average) at 30% of the locations and exceeded WHO indoor air quality standards (50 $\mu g/m^3$, 24-h average) at 92.5% of the locations. The only locations where the concentrations were within WHO limits were the air-conditioned

Table 2 Classification of particle fraction [15]

Classification	Definition	Description
Inhalable fraction	Mass fraction of the particle which is inhaled by nose and mouth	Particles < 100 μm
Thoracic fraction	Mass fraction of the particle that passes through the larynx	Median of the particle size: 11.64 μm Geometric standard deviation: 1.5 μm. 50% of particles with aero diameter of 10 μm
Respirable fraction	Mass fraction that reaches the alveoli	Median of the particle size: 4.25 μm Geometric standard deviation: 1.5 μm. 50% of particles with aero diameter of 4 μm

rooms where the IPM concentrations were 42.4 and 48 $\mu g/m^3$. The highest PM_{10} concentration was found in the concrete laboratory where the level exceeded 10 times the National Ambient Air Quality Standard. The levels of particulate matter at the outdoor sampling stations were less than the indoor lab space except for sampling locations OP6 and OP9 where construction works were going on. The level of IPM_{10} was higher in the structural and concrete laboratory may be because of improper ventilation, dumping of construction and demolition waste, cement and sand particles. Faculties and scholars whose cabin is within the laboratory were highly exposed to IPM_{10}, which in turn will have serious health effects. Details of the sampling locations and the corresponding sampling IDs are shown in Table 3 and Fig. 3.

Table 3 Sampling location description

SP 1: Faculty cabin 1	SP 11: Dumping yard	SP 21: Classroom 1	SP 31: Auditorium
SP 2: Faculty cabin 2	SP 12: Structures lab	SP 22: Classroom 2	SP 32: Canteen
SP 3: Envi. research lab	SP 13: Strength of materials lab	SP 23: Scholar room 4 (AC)	OP1: Near SP3–7
SP 4: Scholar room 1	SP 14: Scholar room 2	SP 24: Scholar room 5	OP2: Near SP 26, 27
SP 5: Envi. lab room 1	SP 15: Scholar room 3	SP 25: Center for transportation	OP3: Near SP 30
SP 6: Envi. lab room 2	SP 16: Conference room	SP 26: Geotechnical lab	OP4: Road 1
SP 7: Lab assistant room	SP 17: Office room 1	SP 27: Lab assistant room	OP5: Road 2
SP 8: GC/MS and AAS room	SP 18: Office room 2	Sp 28: Transportation lab	OP6: Road 3
SP 9: Faculty cabin 3	SP 19: HOD cabin	SP 29: Offshore lab	OP7: Road 4
SP 10: Concrete lab	SP 20: Faculty cabin 4	SP 30: Scholar room 6	OP8: Road 5

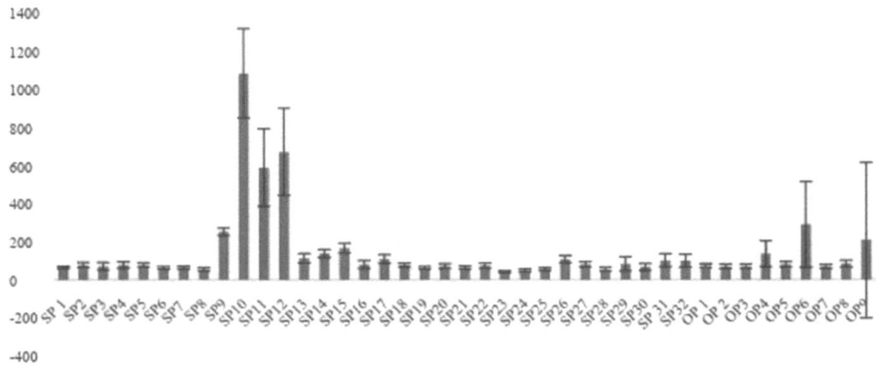

Fig. 3 PM_{10} level

$IPM_{2.5}$ concentration exceeded Indian NAAQS (60 $\mu g/m^3$, 24-h average) at 20% of the locations, and it exceeded WHO standards (25 $\mu g/m^3$, 24-h average) at all the sampling locations. As in the case of PM_{10}, the highest concentration of $IPM_{2.5}$ was observed in the concrete laboratory. Since IPM_1 does not have any standards for ambient air quality and indoor air quality, the values cannot be compared with any of the standards. But the concentrations were found to follow a slightly different trend when compared with that of IPM_{10} and $IPM_{2.5}$. The highest concentration was found in dumping yard of the structural engineering laboratory (Figs. 4 and 5).

Since the instrument has 31 different size channels, it is directly possible to obtain data under different classes, viz. inhalable fraction, thoracic fraction and respirable fraction. The data sets are tabulated in Table 4. As the instrument measures particles of size 30 μm or less, the inhalable fraction shown does not include particle size above 30 μm. On an average, 78% of the inhalable fraction enters the thoracic region, and 51% of the inhalable fraction enters the alveoli region which will give rise to acute and chronic respiratory illness. The transport of particulate matter in the respiratory system of human beings is demonstrated in Fig. 6.

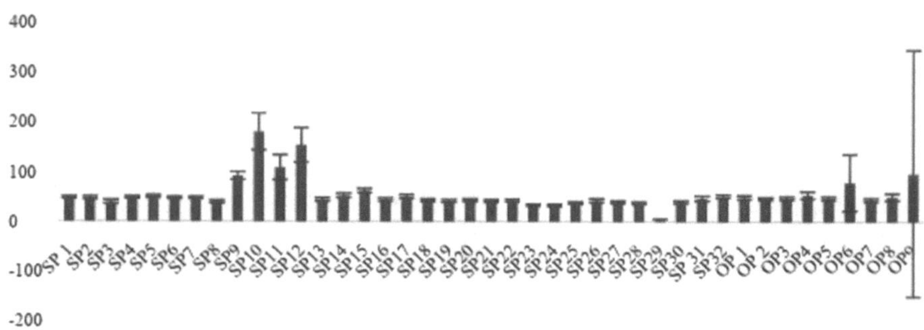

Fig. 4 $IPM_{2.5}$ level

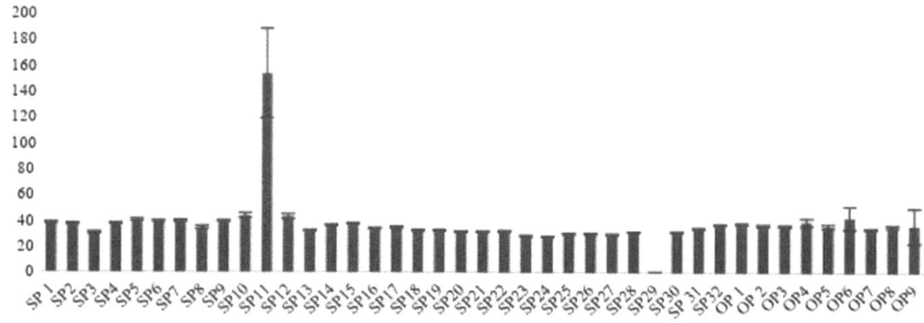

Fig. 5 IPM_1 level

Table 4 Level of IPM inhalable, thoracic and respirable fraction (μg/m^3)

Location	Inhalable fraction mean	Standard deviation	Thoracic fraction mean	Standard deviation	Respirable fraction mean	Standard deviation
SP 1	77.7	37.9	66.7	6.1	53.4	2
SP 2	103.3	31.6	79.4	11.4	54.4	3.4
SP 3	112.5	62	74.3	20.5	45.1	4.9
SP 4	99.1	47.6	79.4	18.6	54.5	2.6
SP 5	105.7	110.2	78.7	11.7	56.6	3.1
SP 6	74.9	28.5	63.1	7.1	51.4	2.3
SP 7	70.9	26.6	63.3	7.8	51.3	2.5
SP 8	63.9	16.8	55.3	8.3	42.6	3.3
SP 9	277.7	57.5	247.5	19.1	139.3	9.7
SP 10	1181.7	264	1023.9	214.7	383.3	69.5
SP 11	727.2	279.7	564.4	192.9	210.6	55.5
SP 12	809.6	355	655.4	225.1	285.5	67.4
SP 13	147	55.8	113	27.1	58	6.2
SP 14	177.1	46.7	136.9	20.1	70.2	5.1
SP 15	261.2	244.9	165.8	27.3	84.3	5.4
SP 16	118.7	69.1	82.7	24	50.8	4.8
SP 17	166.2	143.8	110.1	22.8	63.9	5
SP 18	146.3	212.7	79	11.5	51	3.1
SP 19	66.2	10.9	62	6.6	46.9	3
SP 20	88.9	35.6	72.6	13.5	50.5	3.4
SP 21	69.7	16.1	64.2	8	47.7	3.1

(continued)

Table 4 (continued)

Location	Inhalable fraction mean	Standard deviation	Thoracic fraction mean	Standard deviation	Respirable fraction mean	Standard deviation
SP 22	109.7	88.8	74.4	14.1	48.8	3.1
SP 23	56.1	33.4	44.1	6.7	35.5	1.8
SP 24	57.4	24.7	48.2	8.1	36.2	2.2
SP 25	85.4	111.2	57.5	8.2	42.5	2.3
SP 26	144.5	80	106	20.4	55.9	5
SP 27	98	26.1	80.1	11.7	48.6	3.4
SP 28	65.7	26.7	56.8	11.4	42.2	2.8
SP 29	121.5	94.4	82.7	35.3	48.5	9.4
SP 30	92.7	48.3	68.6	17.6	45.5	4.1

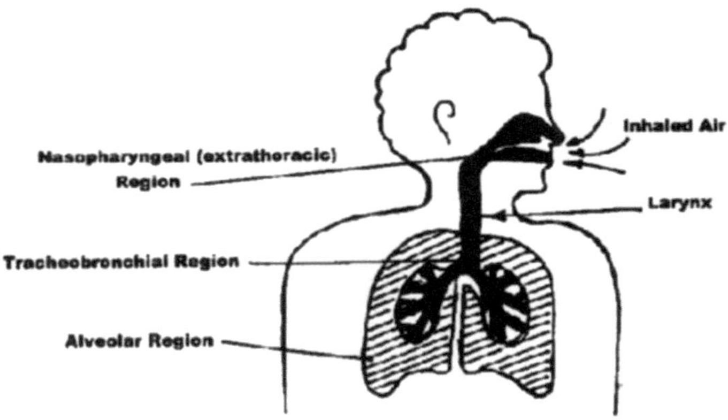

Fig. 6 Schematic representation of the human respiratory tract (reproduced from [16] with permission)

4 Conclusion

Indoor particulate matter of size 10, 2.5 and 1 μm was monitored in laboratories, classrooms and faculty cabin of Civil Engineering Department. IPM concentration in all laboratories exceeded the standards specified by WHO on air quality, and scholars working in structural engineering and concrete lab were exposed to 10 times more pollution when compared with standards prescribed by Indian NAAQS. The level of pollution in the outdoor environment was less when compared to the level of indoor pollution in most of the sampling points. Frequent cleaning and disposal of waste are highly recommended to reduce the pollution level. Moreover, all the scholars should be advised to wear a nose mask when working in the lab to reduce the exposure to IPM.

References

1. Han C, Lim Y, Jung K, Hong Y (2017) Association between ambient particulate matter and disorders of vestibular function. Environ Res 155:242–248. https://doi.org/10.1016/j.envres.2017.02.013
2. Mutlu E, Comba I, Cho T, Engen P, Yazıcı C, Soberanes S et al (2018) Inhalational exposure to particulate matter air pollution alters the composition of the gut microbiome. Environ Pollut 240:817–830. https://doi.org/10.1016/j.envpol.2018.04.130
3. Jamshidi HR (2018) Risk assessment of particulate matters in a dentistry school using fuzzy inference systems. Measurement 257–263
4. Wu JZ (2018) Effects of particulate matter on allergic respiratory diseases. Chronic Dis Transl Med 4(2):95–102
5. Yang T, Liu W (2018) Does air pollution affect public health and health inequality? Empirical evidence from China. J Clean Prod 203:43–52. https://doi.org/10.1016/j.jclepro.2018.08.242
6. Lee SC (2002) Inter-comparison of air pollutant concentrations in different indoor environments in Hong Kong. Atmos Environ 36(12):1929–1940
7. Oeder S (2012) Toxicity and elemental composition of particulate matter from outdoor and indoor air of elementary schools in Munich. Germany Indoor Air 22(2):148–158
8. Chithra VS, Shiva Nagendra SM (2015) Characterizing and predicting coarse and fine particulates in classrooms located close to an urban roadway. J Air Waste Manag Assoc 64(8):945–956
9. Elbayoumi M (2013) Spatial and seasonal variation of particulate matter (PM 10 and PM 2. 5) in Middle Eastern classrooms. Atmos Environ 80:389–397
10. Daisey JM (2003) Indoor air quality, ventilation and health symptoms in schools: an analysis of existing information. Indoor Air 13:53–64
11. BW Business world, Retrieved from http://www.businessworld.in/article/1-3-Million-Deaths-Every-Year-In-India-Due-To-Indoor-Air-Pollution/09-09-2017-125739/. Last Accessed 2/5/2019
12. Apte MG (2000) Indoor carbon dioxide concentrations and SBS in office workers. Proc Healthy Build 1:133–138
13. Goldman M (1978) Industrial bronchitis. Br J Ind Med 35:285–291
14. European Norms (EN-481, 1993), Workplace Atmospheres - Size Fraction Definitions for Measurement of Airborne Particles. Retrieved from https://standards.globalspec.com/std/969024/EN%20481. Last Accessed 2/5/2019
15. The Ultimate New Model 11-R technical datasheet, Retrieved from http://www.envitech-bohemia.cz/files/008-indoor/grimm/01-mini-las/mini-las-en.pdf. Last Accessed 2/5/2019
16. World Health Organization Geneva (1999), Hazard Prevention and Control in the Work Environment: Airborne Dust. WHO/SDE/OEH/99.14. Retrieved from https://www.who.int/occupational_health/publications/en/oehairbornedust.pdf. Last Accessed 2/5/2019

Assessment of Indoor Environmental Quality and Impacts on Occupants: Case Study of MNIT Jaipur

Nivedita Kaul and Khushbu Parik

Abstract Indoor environmental quality (IEQ) is one of the major factors for producing a favourable and productive environment in any educational building. Various studies show the effect and consequences of poor IEQ on grasping capability of students. Today institute buildings are expected to provide enhanced IEQ, higher occupant satisfaction and less risks of occupant health. With reference to various results available, a study has been conducted in Malaviya National Institute of Technology, Jaipur, Rajasthan, for analysing the IEQ conditions in the institute's building. The study is conducted in two parts, first includes quantitative analysis of various IEQ parameters like temperature, relative humidity, noise, air ventilation, particulate matter, volatile organic carbon, light intensity and CO_2 concentration. These parameters are further analysed as per the ASHRAE and various other standards. The second one includes survey for determining the student's perception towards the current environment and comfort in the building. The study included various classrooms, offices and working areas of the institute to analyse the difference in concentration of parameters as per the usage and occupancy of the area. In this research, it is evident that level of satisfaction for majority of users in the academic building of MNIT is moderate. The reason accounted for the discomfort of various students may be indoor noise generated by some of the ACs, improper ventilation in summer season and construction activities in the vicinity. A few actions should be taken into consideration to produce a good IEQ environment at the academic building and increasing the productivity of the students and workers of the institute.

Keywords IEQ · Occupant health

N. Kaul (✉) · K. Parik
Department of Civil Engineering, Malaviya National Institute of Technology, Jaipur, Rajasthan, India
e-mail: kaulmnitj@gmail.com

A. Sharma et al. (eds.), *Indoor Environmental Quality*, Lecture Notes in Civil Engineering 60,
https://doi.org/10.1007/978-981-15-1334-3_13

1 Introduction

The effects of IEQ on cognitive performance on the students are largely under perceived in India. Poor IEQ can give negative impact to facilities, buildings and occupants, thus affecting the teaching and learning process adversely. Thermal comfort levels and indoor air quality (IAQ) play a crucial role in producing a favourable environment that supports educational and health outcomes. There is enough evidence in the literature to support the fact that indoor pollutants and thermal conditions in educational institutes may influence student's performance or attendance. Additionally, there is significant evidence that institute exposure to various indoor air pollutants (IAP) may have health and comfort implications on the students, which may impair performance indirectly. For example, exposure to noise, excessive heat and poor lighting conditions may affect the concentration of student and their grasping power of the subject delivered. Indoor environmental quality (IEQ) must become an important component of providing quality education in our learning institutes. Today institute buildings are expected to provide enhanced IEQ, higher occupant satisfaction and less risks of occupant health. A study has been undertaken in various institutional areas of MNIT, Jaipur, to evaluate the existing conditions of IEQ and their effect on the performance and well-being of the students.

2 Methodology

Malaviya National Institute of Technology (MNIT) is located in Malviya Nagar, Jaipur, Rajasthan. There are hospitals, business centre, residential and commercial areas located adjacent to the main campus, i.e. within 1 km radius. The study focuses on various indoor micro-environments including classrooms, laboratories and offices of MNIT. Locations (13) have been considered for data collection.

The present study consists of two parts. First includes measurement of IEQ parameters on selected locations to determine the quantitative data in the present conditions. The second part includes survey for understanding the perception of respondents on the present IEQ conditions of the building. The major IEQ parameters that have been considered in the present study are: temperature, relative humidity, noise, air ventilation, particulate matter, volatile organic compounds (VOCs), light intensity and CO_2 concentration. Various instruments used for the data collection are temperature and humidity metre (FLUKE 971), air quality monitor (RAVE), lux metre (MECO), anemometre and sound level metre. The study sites have been selected on the basis of size of the room, purpose of the room (lecture theatre/laboratory/staff cabin etc.), ventilation conditions (natural/air-conditioned), age of the building (new/old building), possibility of persistent IAQ problems due to proximity to toilets, ongoing construction activities etc. Measurements have been taken for a duration of 1 h at each location selected for study in summer season. In each classroom, the measurement has been taken at different times of the day and at different locations to record more precise data.

For studying the occupant's perception, questionnaire survey forms have been distributed to the students in classrooms. Questionnaire has been developed based on research objectives to collect research data through various Likert scale questions and some opinion-based questions, on the following parameters

Thermal Comfort	
Rating	Comfort Level
1	Uncomfortable
2	Poor
3	Moderate
4	Good
5	Comfortable

Odour Intensity	
Rating	Intensity
1	No Odour
2	Very Weak
3	Weak
4	Distinct
5	Strong
6	Very Strong

Background Noise	
Rating	Noise Level
1	Very high
2	High
3	Moderate
4	Low
5	Silent

Lighting conditions	
Sufficient	
Not sufficient	

Audibility Problem	
Yes	
No	

Dust Problem	
Yes	
No	

Stuffy feeling	
Yes	
No	

Cleanliness	
Always	
Frequently	
Never	

3 Results and Discussion

The data collected is analysed as per American Society for Heating Refrigeration and Air Conditioning Engineers (ASHRAE), Illuminating Engineering Society of North America (IESNA), IAQ index, World Health Organisation (WHO) and Indian Standard (IS) standards. Relative humidity (RH) of the atmosphere is an important parameter determining thermal comfort. As per ASHRAE, RH should be between 30 and 65%. In the present study, the relative humidity at six locations is within the range. The VOCs have been observed to be within permissible limit of 1 ppm at all the locations. Though the lighting intensity at most of the locations has been observed to be less than 400 lx (IESNA standard), still the perception towards lighting condition is satisfactory which may be due to the natural light from the windows. The carbon dioxide concentration is in good category at maximum of the places as per ASHRAE. However, a high concentration of CO_2 (1990 ppm) is observed at one of the locations which may be due to improper ventilation and higher occupancy at the time of measurement. PM_{10} and $PM_{2.5}$ have been observed within the satisfactory limit except at two locations. The reason for higher concentration may be due to ongoing construction activity near the location. The variation in temperature should range between 22.8 and 26.1 °C in summer and 20 and 23.6 °C in winters as per the ASHRAE standards. However, the observed temperature is higher than the prescribed limits for summer seasons.

In order to analyse the perception of IEQ elements at MNIT, Jaipur, a survey has been carried out at the locations under consideration by distributing 320 questionnaires to the occupants of the study sites. Responses (232) were completed and have been analysed for the perception of various IEQ parameters. The perception of the noise comfort belongs to subjective feeling, and surrounding noise may disturb the people even if the noise level does not exceed the relevant standards. The survey results show that 8.29% students have reported that classrooms are subject to very noisy conditions, 9.46% have reported high noise, and 31.36% have reported moderate noisy conditions, while 40.23% and 10.65% have reported low and silent level of noise, respectively.

Lighting can change the mood of people and affects their psychology. In this study, 74.14% students have reported sufficient lighting conditions while 25.85% have problems with lighting conditions and reported it unsatisfactory. 43.54% students have reported the feeling of stuffiness in the classrooms. This problem may be due to improper ventilation in the class. The survey has been done in summer season, and the number of occupants is high which also contributes to comfort issues at the location. Dust has not been reported a problem by majority of respondents (66.2%) due to frequent cleaning in the classrooms. Audibility condition in the classrooms has been reported to be unsatisfactory by 56% of students which may be due to the construction activities at the time and other indoor noises. For odour intensity, perception varied from ‘no’ (39.58%) to ‘weak odour’ (19.27%). Some of the odour problems can be due to sensitive to a particular cleaning agent. In spite of high temperature than the suggested limits, 46.66% students reported moderate thermal comfort level in summers, while 53.12 reported good comfort level in winter season.

4 Conclusion

The study shows that the administration must give substantial importance to IEQ parameters and provide adequate thermal comfort, proper audibility conditions, ventilation and efficient lighting so as to ensure good teaching and learning condition in educational institutes. In this research, it is evident that the level of satisfaction for majority of users in the academic building of MNIT is moderate. A few actions should be taken into consideration to produce a good IEQ environment at the academic building. Improper maintenance will affect the building's environment quality, and it highly influences productivity and well-being of the facility management. Priority should be given on the maintenances aspect as this will affect the core functioning of the institute.

Classroom Ventilation and Its Impact on Concentration and Performance of Students: Evidences from Air-Conditioned and Naturally Ventilated Schools of Delhi

Pratima Singh, Renu Arora and Radha Goyal

Abstract Carbon dioxide (CO_2) concentration acts as an important indicator of indoor air quality inside school premises. CO_2 concentration above 1000 ppm inside a building is an indicator of insufficient ventilation which may cause health complaints to the occupants and may directly or indirectly impair concentration and performance of the students. Realizing the need for research in this area, an attempt was made to investigate the relationship between classroom ventilation and Concentration Performance (CP) of school children. In this paper, the classroom ventilation was evaluated through the concentration of CO_2 inside air-conditioned (AC) and naturally ventilated (NV) urban school buildings located in South Delhi. The results indicated the evident problem of elevated concentrations of CO_2 inside the AC schools which often exceeded the ASHRAE's recommended limit of 1000 ppm. The results from the d2 test reflected that mean CP score of students was higher in the schools with lower CO_2 concentrations than students from schools with higher CO_2 values. However, the differences in their mean were not significantly different (p-value > 0.05). The current research suggested the need for longitudinal study to explore the relationship of CO_2 with students' performance inside the classrooms.

Keywords Indoor air quality · Carbon dioxide · School · Performance · Student · Ventilation

1 Introduction

According to the World Health Organization (WHO), clean air is the basic requirement for human health and well-being [1]. As we normally breathe about 12,000 l of air every day, it is essential for our health to have clean air in our environment [2].

P. Singh (✉) · R. Arora
Institute of Home Economics, University of Delhi, F-4 Hauz Khas Enclave, New Delhi 110016, India
e-mail: pratima.singh@ihe.du.ac.in

R. Goyal
Indian Pollution Control Association, Hargobind Enclave, New Delhi 110092, India

A. Sharma et al. (eds.), *Indoor Environmental Quality*, Lecture Notes in Civil Engineering 60,
https://doi.org/10.1007/978-981-15-1334-3_14

Central Pollution Control Board's (CPCB) Indoor Air Pollution Report (2014) refers indoor air quality (IAQ) to the quality of air inside the buildings as represented by the concentrations of the pollutants and thermal conditions (temperature and relative humidity) that affect the health and performance of the occupants [3]. Along with the pollutants in ambient air, building-related conditions such as building materials and their permeability, air-conditioning and ventilation systems used also affect quality of air inside a space [4].

Understanding the impact of the indoor air on children is important as they spend most of their time indoors (approximately 80–90%), either at school or at home. Within a school, students spend most of their time inside classrooms, engaging in varied activities which require a considerable amount of concentration and attention. Several research studies have indicated that the air quality inside classrooms is often poor and CO_2 concentrations frequently exceed recommended levels prescribed by the American Society of Heating, Refrigerating and Air-Conditioning Engineers (ASHRAE) for indoor environments, which might be due to high student density or poor ventilation strategies adopted in schools [5]. Faulty heating, ventilation, and air-conditioning systems in schools can exacerbate air quality problems inside classrooms leading to 'sick building syndrome' manifestation on children's health in the form of various health issues. If air inside classrooms is deteriorated, it may cause several health effects that may directly impair concentration or memory of students or cause other health effects that indirectly affect their learning.

Carbon dioxide (CO_2) is colourless, odourless, and tasteless gas which is non-flammable. The indoor concentration of CO_2 can be a good measure of the ventilation per person inside the space [6]. In most locations, the CO_2 level in outdoor air is reported to be in a range of 350–450 ppm [7, 8]. The higher CO_2 concentrations in the outdoor environment can be due to heavy vehicular traffic, industries, and varied sources of combustion, whereas, the elevated indoor concentration of CO_2 can be a direct result of respiration by building's occupants and inadequate ventilation (Fig. 1).

As classrooms are being densely occupied by a large number of students, they may exhibit elevated levels of CO_2 resulting in drowsiness, lethargy, and a general sense that the air is stale as reported by students and staff in a number of research

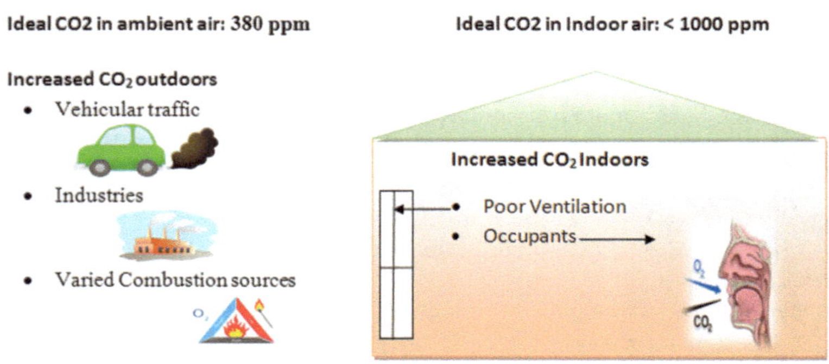

Fig. 1 Factors contributing to CO_2 generation and accumulation inside the space

studies [5]. This general discomfort amongst students may further translate into poor performance inside the classrooms.

Hence, the present research was designed to study the impact of classroom ventilation on the Concentration Performance (CP) of the students studying in air-conditioned (AC) and naturally ventilated (NV) schools located in Delhi. The specific objectives of the research study included:

- To examine the physical space design of the classrooms of selected AC and NV schools.
- To measure the existing indoor CO_2 concentrations in the classrooms of selected schools.
- To analyze the variations in CO_2 concentrations in the AC and NV schools and their relationship with the concentration and performance of the students.

2 Methodology

2.1 Sample Selection

For the purpose of the research, the data was gathered from two AC (schools A and B) and two NV (schools C and D) schools located in South Delhi; which were selected through purposive sampling technique. All the schools were selected from South Delhi for two reasons:

- Limitation of taking all the schools from near proximity of the Delhi Pollution Control Committee (DPCC)'s air quality monitoring station located in RK Puram, Delhi.
- Majority of the AC schools were located in South Delhi.

Three classrooms from each school were selected for CO_2 monitoring based on the permission given by the school authorities. Systematic sampling technique was used for selecting student respondents from the four schools—A, B, C, and D. Informed consent was taken from all the students studying in the selected classrooms and their parents. Finally, a total of 738 students (i.e. 384 in non-winter season and 354 in winter season) were selected from these classrooms as cohorts of this study. The detailed procedure followed to select schools is illustrated in Fig. 2.

2.2 Data Collection

The design features of the classrooms viz., orientation, area, dimensions, ventilation type, number of doors, windows, etc., were studied using a structured checklist and observation tool. The ultrasonic measuring tape 'Bosch GLM 50 Professional' was used for taking measurements of room and window dimensions.

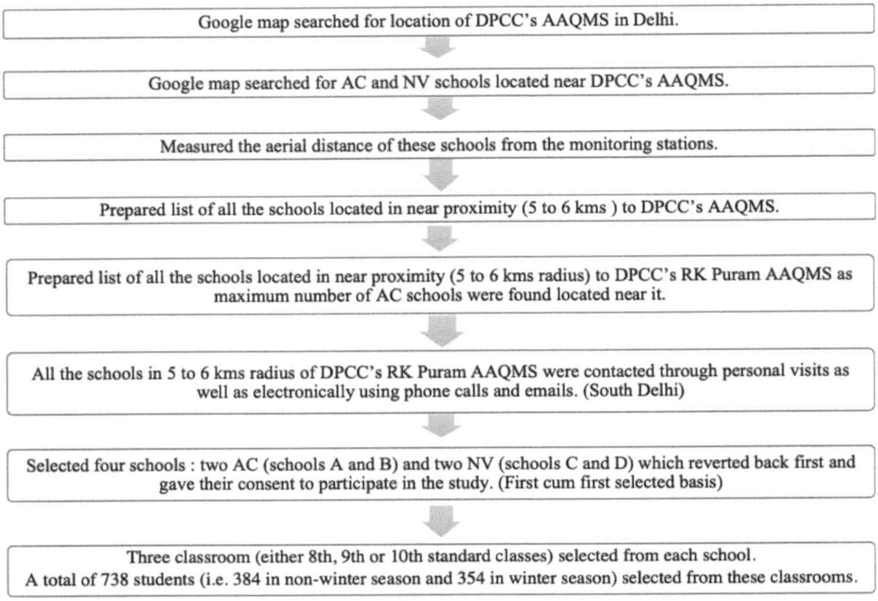

Fig. 2 Procedure followed for sample selection (Note: DPCC—Delhi Pollution Control Committee, AAQMS—Ambient Air Quality Monitoring Station, AC—air-conditioned, NV—naturally ventilated)

The protocol for monitoring of CO_2 was planned from July 2016 till February 2017 (non-winter season: July–October 2016; winter season: November 2016–February 2017) in order to consider seasonal variations. CO_2 monitoring was carried out for 6–7 working hours at continuous 5 min in each classroom on one working day and one non-working day (Saturday) in each school in both the seasons using indoor air quality monitor 'Testo-435' (Tables 1 and 2). The sampling instrument was placed in the centre (highest student density area) of the class at approximately 1 m above the floor level, corresponding to the breathing zone of the occupants. Class activity pattern of the children was also recorded to make valid interpretations of the results.

Further, students' concentration and performance were evaluated through standardized d2 test for speed and accuracy of task completion. The test was created

Table 1 Specifications of indoor air quality monitor 'Testo-435'

Monitoring instrument	Working principle	Range	Accuracy	Measurement type
IAQ Monitor: Testo-435 	Non-dispersive infrared sensor (NDIR)	0 to + 10,000 ppm	± (75 ppm CO_2 ± 3% of mv)	Continuous

Table 2 CO_2 monitoring protocol followed in non-winter (July–October 2016) and winter (November 2016–February 2017) seasons

Type of day	Working days	Non-working days
Number of classrooms monitored in each school	3 classrooms × 4 schools	1 classroom × 4 schools
CO_2 monitoring duration	NV school classrooms: For 6½ h at 5 min interval (7.30 am to 2.00 pm) AC school classrooms: 7½ h (7.30 am to 3.00 pm)	Both AC and NV schools' classrooms: For 6 h at 5 min interval (8:00 am to 2:00 pm)
Total number of days	Non-winter season: 12 days Winter season: 12 days	Non-winter season: four days Winter season: four days

by Brickencamp and Zilmer (1998) and is also known as cancellation task and takes approximately 8–10 min to complete [9]. The Concentration Performance (CP) scores of the students were calculated by subtracting the total incorrect responses from the total correct responses. Descriptive statistics and other statistical analysis parameters were determined through SPSS software and Microsoft Excel.

3 Results and Discussion

3.1 Space Design Observations

Design features of the classrooms of all the selected schools were studied in detail using structured checklist. It was observed that classrooms of AC schools had non-operable compact windows (Fig. 3) in order to reduce heat exchange for desired thermal insulation, which limited the supply of fresh air inside the classrooms. The doors of the classrooms were usually kept closed throughout the day resulting in

(a) School A classroom (b) School B classroom

Fig. 3 Non-operable compact windows in classrooms of AC schools

(a) School C classroom (b) School D classroom

Fig. 4 Operable windows in classrooms of NV schools

CO_2 build-up inside the classrooms. It was also observed that though the school with heating, ventilation, and air-conditioning (HVAC) system, i.e. school B had provision for exhaust fans for removal of excess CO_2 and stale air, they were never used pre-, during or post-occupancy. Whereas in NV schools, windows and doors were kept open most of the time which helped in reducing the CO_2 concentration in the classes of NV schools (Fig. 4).

3.2 *Average Concentrations of CO_2*

The mean CO_2 on working days for each school is a representation of average of CO_2 values obtained from the three different classrooms of each school on three different working days. The given Fig. 5 represents the non-winter and winter trends for concentrations of CO_2 in schools A, B, C, and D.

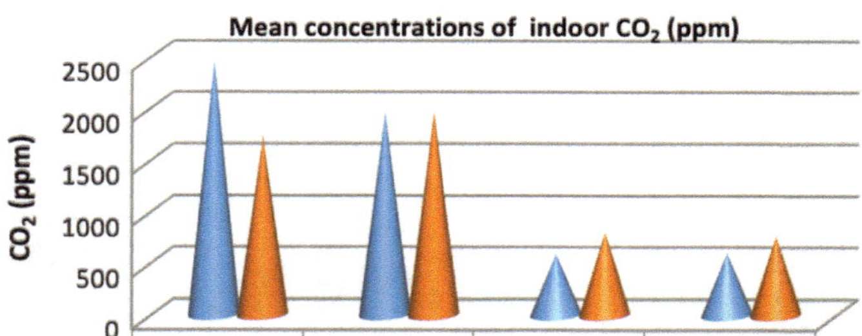

	A	B	C	D
Non-Winter Season	2461.1	1954.7	581.9	593.5
Winter season	1726.2	1952.1	791.5	751.4

Fig. 5 School-wise mean concentrations of indoor CO_2 on working days in non-winter and winter seasons

In non-winter season, the mean values of CO_2 on working days in AC schools were 2479.1 ppm ± 1146.0 ppm (school A) and 1954.7 ppm ± 852.2 ppm (school B) and in NV schools were 581.9 ppm ± 152.6 ppm (school C) and 593.5 ppm ± 141.3 ppm (school D). Whereas in winter season, the mean values of CO_2 on working days in AC schools were 1726.2 ppm ± 861.6 ppm (school A) and 1952.1 ppm ± 695.0 ppm (school B) and in NV schools were 791.5 ppm ± 309.0 ppm (school C) and 751.4 ppm ± 224.0 ppm (school D).The results indicated that the concentrations of CO_2 varied between schools based on the type of ventilation systems used, closing and opening of windows, and classroom occupancy.

Further, it was observed that the AC schools recorded mean CO_2 (i.e. 2207.9 and 1839.13 ppm) much above the limit of 1000 ppm prescribed by ASHRAE standards [10]. However, in both the seasons, the mean concentrations of CO_2 in NV schools were lower than in AC schools and were within recommended limits (587.7–771.5 ppm) (Fig. 6). The open windows and doors of NV classrooms led to higher air exchange rate resulting in lower accumulation of CO_2, whereas closed doors and windows in AC classrooms resulted in higher accumulation of CO_2 generated as a result of occupants' respiration. Research studies by Jurado et al. [11] and Yang Razali et al. [12] also linked high concentrations of CO_2 with human respiration and different ventilation systems used in the classrooms.

The results further highlighted that the mean values of CO_2 on non-working days were lower than on working days in all schools in both the seasons (Table 3), as the sources of CO_2 generation were negligible on weekends due to no classroom occupancy. In a similar research study by Yang Razali et al., a high concentration

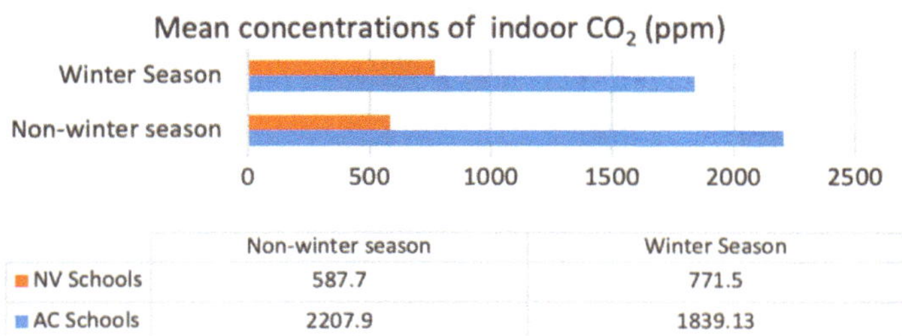

Fig. 6 Mean concentrations of indoor CO_2 in AC and NV schools in non-winter and winter seasons

Table 3 Comparison of CO_2 on working and non-working days

Mean CO_2 concentration (in ppm)								
Schools	A		B		C		D	
Day	WD	NWD	WD	NWD	WD	NWD	WD	NWD
Non-winter season	2461.1	484.2	1954.7	407.2	581.9	462.5	593.5	388.1
Winter season	1726.2	509.6	1952.1	347.8	791.5	351.0	751.4	360.0

Note A and B: AC schools, C and D: NV schools, WD: working day, NWD: non-working day

of CO_2 was found to be closely related to human respiration and closed windows and doors [12]. Similarly, in the current research, values of CO_2 were considerably lower on non-working days due to no human respiration in classrooms.

3.3 Effect of Classroom Occupancy and Activities on CO_2 Concentrations

The concentration of CO_2 is considerably higher during the full occupancy periods due to human respiration than when classrooms are vacant or have lesser occupancy [12]. To study the effect of classroom occupancy on the variations of CO_2, data was collected at 5 min interval for a complete working day from each classroom.

School A: The high values of CO_2 indicated poor ventilation in all the classrooms of school A. All the windows of classrooms of school A were non-operable except for one small emergency window. It was observed that whenever emergency windows or doors were opened or when the classrooms were empty, CO_2 levels showed a declining trend in both the seasons in the classrooms of school A.

Non-winter season: As the day progressed, increased room occupancy resulted in an upward trend in CO_2 values. CO_2 started to accumulate inside the classrooms during full occupancy hours. As the students dispersed at 2:20 pm, after which the classrooms were empty, CO_2 started showing a declining trend in all the classrooms (Fig. 7a).

Winter season: In winter season, classrooms experienced higher CO_2 levels (often exceeding 1000 ppm) in early morning hours due to previous days CO_2 build-up indoors, thereby, exposing students to unhealthy indoor conditions (Fig. 7b). The doors and one small emergency window were sometimes opened for occupants' comfort in winter season. Hence, the mean CO_2 in winter season was lower than in non-winter season in school A.

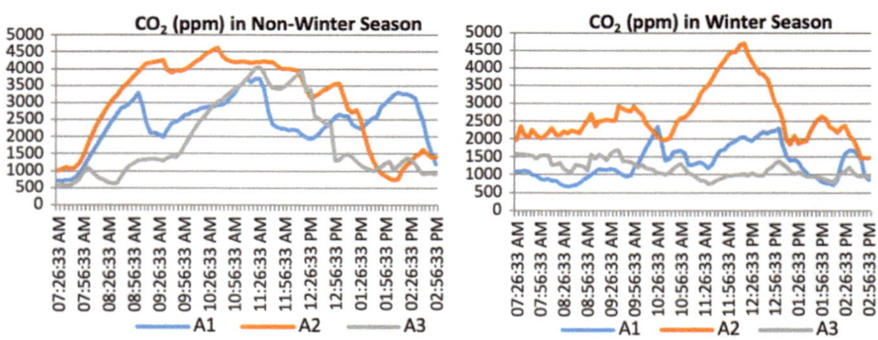

(a) Non-winter Season (b) Winter Season

Fig. 7 Variations in concentrations of CO_2 during working days in classrooms A1, A2, and A3 in **a** non-winter season and **b** winter season

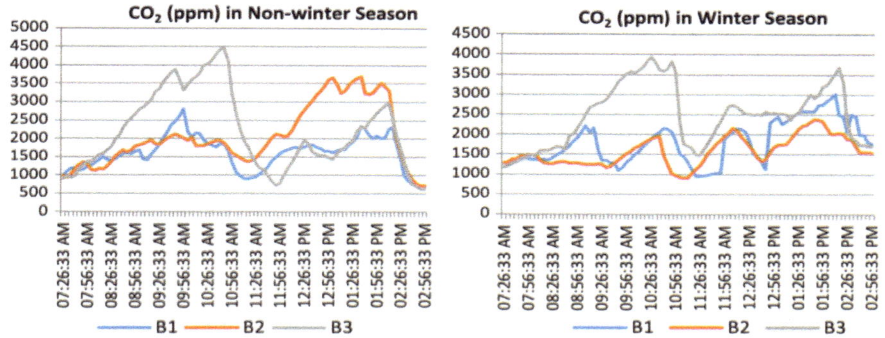

(a) Non-winter Season (b) Winter Season

Fig. 8 Variations in concentrations of CO_2 during working days in classrooms B1, B2, and B3 in **a** non-winter season and **b** winter season

School B: The windows of classrooms of school B were non-operable and hence, the classrooms were completely dependent on HVAC system for ventilation. Therefore, all the classrooms showed almost similar trends of CO_2 in both the seasons (Fig. 8a, b). The values of CO_2 were low during morning hours in all the three classrooms, since CO_2 was in an early building upstage. As the day progressed and students started to settle, CO_2 values started to gradually increase. A few sharp peaks were also observed in CO_2 values. These were the time periods when classroom occupancy was full and doors were closed. These peaks slowly declined when room occupancy was less or classrooms were empty. On some occasions, in spite of the room occupancy being same, opening of door for a longer duration resulted in a drop in CO_2 readings for some time, which again shoot up with the closing of doors.

School C: Classroom of school C relied mainly on fans and windows for ventilation.

Non-Winter Season: Throughout the day, doors and windows were kept open in all the classrooms for circulation of fresh air. This resulted in almost consistent values of CO_2 for the entire duration of monitoring. Also, the CO_2 values did not exceed 1000 ppm in any of the classrooms in non-winter season.

Winter Season: Relative to non-winter season, CO_2 levels in winter season were higher and few peaks were also observed because of few incidences of closing of doors and windows. In winter season, CO_2 values occasionally crossed 1000 ppm on certain time periods. However, the mean values of CO_2 were less than 1000 ppm in both seasons (Fig. 9a, b).

It was observed that though the ventilation was adequate in both the seasons in school C classrooms, the mean concentrations of CO_2 in the winter season were greater than of non-winter season. The reason for the same was intermittent closing of doors and windows in winter season to protect occupants from cool wind waves. Closing of windows for a short period of time was also linked with fluctuations in CO_2 readings.

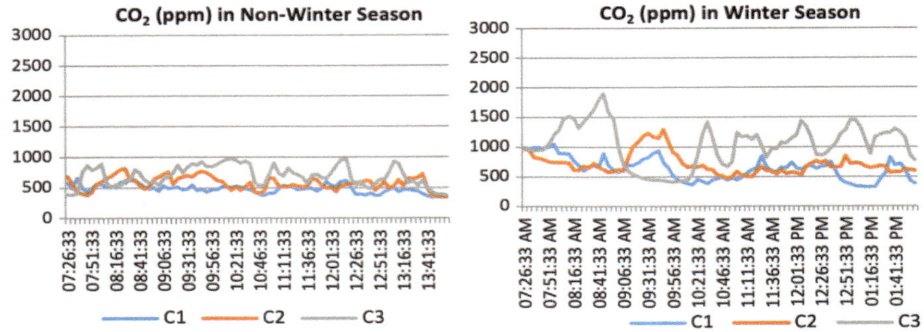

(a) Non-winter Season (b) Winter Season

Fig. 9 Variations in concentrations of CO_2 during working days in classrooms C1, C2, and C3 in **a** non-winter season and **b** winter season

School D: Similar to school C, the doors and windows were kept open in all the classrooms of school D in non-winter season which resulted in consistent values of CO_2 for the entire monitoring period in non-winter season. Whereas in winter season, the mean values of CO_2 in all the classrooms of school D were more than in non-winter season (Fig. 10a, b) because of intermittent closing of doors and windows for occupants' thermal comfort.

Overall, seasonal variations were observed in mean values of CO_2 in all the classrooms (except school B). CO_2 concentrations were directly related to low ventilation rates in densely populated classrooms.

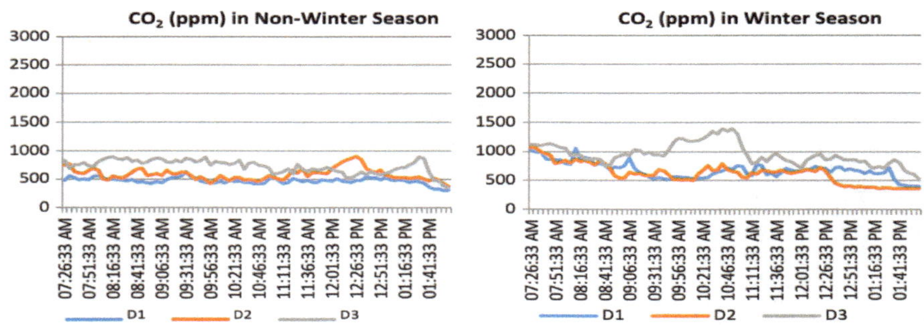

(a) Non-winter Season (b) Winter Season

Fig. 10 Variations in concentrations of CO_2 during working days in classrooms C1, C2, and C3 in **a** non-winter season and **b** winter season

3.4 *Effect of Classroom Ventilation (CO_2 Concentrations) on Concentration and Performance of Students*

When comparisons were made between the mean Concentration Performance (CP) scores of students of the NV and AC schools, the results highlighted that the mean score of the students of NV schools (197.06) was higher than of students of AC schools (189.13) (Fig. 11). However, the differences in their mean were not significantly different (p-value > 0.05) (Table 4). As the students of AC schools were exposed to higher CO_2 concentrations often exceeding 1000 ppm, the effect of elevated CO_2 was evidently reflected on their lower CP scores compared to the students of NV schools who rarely experienced elevated CO_2 inside their classrooms. Previous research studies also showed similar results, wherein, students' task speed increased with the increase in ventilation rate and decrease in CO_2 levels [13]. Their study provided strong evidence of linkage between low ventilation rates to reduced students' attention and vigilance and negative effects on memory and concentration.

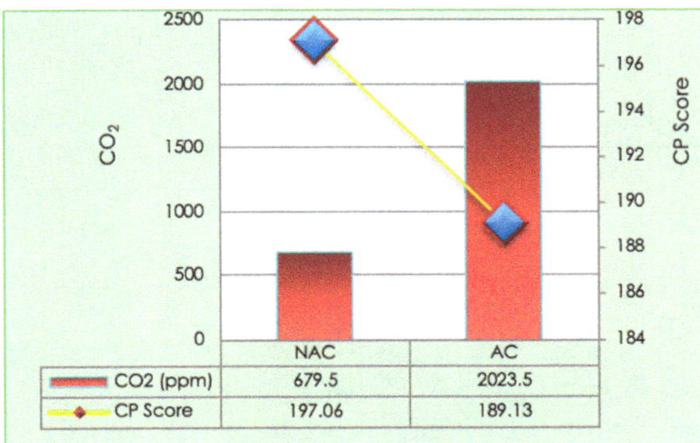

Fig. 11 Comparison of mean CP scores and mean CO_2 concentrations of AC and NV schools

Table 4 Mean CP scores of the students

School/season	Mean	Std. deviation	t	p-value
NV (n-415)	197.06	54	1.763	0.078*
AC (n-323)	189.13	65		
Total (n-738)	193.09	61		

Note Independent sample t test, p-value is significant at 0.05*

4 Conclusion

The study decisively indicated that the inward flow of outside fresh air and occupancy of a room played important roles on the concentrations of CO_2 inside the classrooms. Open doors and windows helped in diluting accumulated CO_2; thus, helping in improving the quality of air indoors. The elevated indoor CO_2 concentrations in AC schools may indicate inadequate ventilation per occupant, which may severely affect students' health and performance in schools.

Based on the observations of the current study, the AC schools are advised to increase ventilation rate so that CO_2 levels can be controlled. The CO_2 sensors must be installed inside the classrooms and exhaust fans can be used whenever CO_2 concentrations go beyond the permissible limits. Length of breaks should be adequate so that the accumulated CO_2 can be removed via adequate ventilation techniques during break periods. The doors and windows of the classrooms must also be opened well in advance before and at the end of the working day for re-circulating fresh air from outside.

Acknowledgements We acknowledge technical help and support provided by Indian Pollution Control Association during data collection for this research.

Ethical Approval The research design and related protocols for the study have been approved by the Institutional Ethical Committee, Institute of Home Economics, University of Delhi on 1 June 2016.

References

1. WHO Guidelines for indoor air quality-selected pollutants. Denmark, 978-9-28900-213-4 (2010)
2. ISHRAE (2015) Indian society of heating refrigerating and air conditioning engineers position paper on indoor environmental quality. http://ishrae.in/pdf/IEQ-2015.pdf. Last accessed on 05 Apr 2017
3. CPCB (2014) Indoor air pollution: monitoring guidelines. http://www.cpcb.nic.in/Proto-Ind_AirPollution_June2014.pdf. Last accessed on 15 June 2017
4. Mudakavi JR (2010) Indoor air pollution. Air principles and practices of air pollution control and analysis. I. K. International Pvt Ltd., pp 369–370
5. Daisey JM, Angell WJ, Apte MG (2003) Indoor air quality, ventilation and health symptoms in schools: an analysis of existing information. Indoor Air 13(1):53–64
6. Hou Y, Liu J, Li J (2015) Investigation of indoor air quality in primary school classrooms. Proc Eng 121:830–837. www.sciencedirect.com. Last accessed on 02 Apr 2017
7. Seppanen OA, Fisk WJ (2003) Summary of human responses to ventilation. Indoor Air 14:102–118
8. Hussin M, Ismail MR, Ahmad MS (2017) Air-conditioned university laboratories: comparing CO_2 measurement for centralized and split-unit systems. J King Saud Univ—Eng Sci 29(2):191–201
9. Brickencamp and Zilmer (1998) d2 Test of Attention, 1st edn. Hogrefe & Huber
10. American Society of Heating, Refrigerating, and Air Conditioning Engineers (ASHRAE). ASHRAE Standard 62(1992) Ventilation for Acceptable Indoor Air Quality. Atlanta, GA

11. Jurado SR, Bankoff ADO, Sanchez A (2014) Indoor air quality in Brazilian universities. Int J Environ Res Public Health 11:7081–7093. https://doi.org/10.3390/ijerph110707081
12. Yang Razali NY, Latif MT, Dominick D, Mohamad N, Sulaiman FR, Srithawirat T (2015) Concentration of particulate matter, CO and CO_2 in selected schools in Malaysia. Build Environ Sci Direct 87:108–116 (Elsevier)
13. Wargocki P, Wyon D (2006) Effects of HVAC on student performance. ASHRAE J 48:22–28

Indoor Air Quality and Thermal Comfort in Green Building: A Study for Measurement, Problem and Solution Strategies

Praveen Babu and Gourav Suthar

Abstract Indoor air quality (IAQ) has become a global health issue due to rapid urbanization resulting into construction of air tight high rise buildings affecting the indoor environment. Majority of the people spend an average of 87% of their time within enclosed buildings; living, working and studying. Various types of pollutants are present indoor that are emitted either by natural or anthropogenic activities thus adversely affecting the health of occupants within the building. Major sources of indoor air pollutants are occupants/building users, their activity, appliances, building materials and infiltration of pollutants from outdoors. The concept of sustainability of green buildings includes a wide range of socio-economic and environmental problem. The increase in concentration of indoor pollutant has decreases the indoor air quality and the inadequate design of the green building may cause lower thermal comfort. The pollutant released from the building material and furnishings causes more harm to the occupants but still, there is a scope of improving the IAQ and thermal comfort by implementing the proper natural or mechanical ventilation, air cleaning, proper design of the building material, etc. Through a literature review, a study has been carried out by elaborating the major causes of poor IAQ and thermal comfort. This study also depicts the measurement techniques and strategies for the same. It was concluded from the study that indoor environment could be improved by implementing proper methods and system by following the standards and codes from the different organization before designing the green building, i.e. American Society of Heating, Refrigerating and Air Conditioning Engineers (ASHRAE) and Leadership in Energy and Environmental Design (LEED) standards and many more such available standards.

Keywords Indoor air quality · Ventilation · Green building · Comfort

P. Babu (✉) · G. Suthar
Department of Civil Engineering, Amity University Haryana, Gurugram, India
e-mail: praveenb@ggn.amity.edu

G. Suthar
e-mail: suthargourav7@gmail.com

A. Sharma et al. (eds.), *Indoor Environmental Quality*,
Lecture Notes in Civil Engineering 60,
https://doi.org/10.1007/978-981-15-1334-3_15

1 Introduction

1.1 Indoor Air Quality

Measurement of indoor air quality is the most important factor in the green building since it provides more fresh air than the conventional building by avoiding sources contaminants, which causes air pollution, controls the thermal comfort and improves the maintenance [1, 2]. There are many gases emitted with consumption of building energy causes the change in climate [3]. Indoor air quality (IAQ) has been the main concern for any green building because poor air quality causes severe health problems. Therefore, maintaining a green environment inside the building contributes 7.5% in green building certification [4]. Contact to indoor air pollutants like particulate matter (PM), volatile organic compounds (VOCs), environmental tobacco smoke (ETS) and nitrogen dioxide (NO_2) have been related with the poor health consequences, i.e. asthma, cancer and to overcome from this type of problems, an assessment of IAQ was carried out which reduces numerous indoor exposures and enhanced health results for community shifting from their conventional housing into green buildings [5–7]. Environmental impact studies for green building and its energy consumption were carried out for reducing the severe health effects [8–10]. It was found that there is a relationship between IAQ and the ventilation rate; a good IAQ is attained by satisfactory ventilation with fresh outdoor air. Therefore, keeping the building clean and to make better operations and maintenance, and control on source of indoor air pollution by the selection of suitable building material, isolation of source, etc., is essential [10]. ASHRAE, 2009 is the standard for document determine the indoor air pollution, document for contractors, designers, and others to gain data for achieving a good IAQ. Indoor air difficulties in new houses were mostly instigated from building materials, household products and furnishings [11, 12]. It was observed that advanced mechanical and ventilating systems had been used to increase flow of air and decrease inhabitant interaction with the airborne infectious agents [13].

1.2 Thermal Comfort in the Indoor Environment

The construction sector has changed its trend from conventional sustainable construction [14]. In building, the major cause was found behind the poor air quality and thermal comfort is the inadequate ventilation, and so, a traditional method of the natural ventilation could be applied for the thermal comfort [15]. As per ASHRAE, standard 55 (2013), there are six factors are responsible for change in thermal comfort inside the green building which are air temperature, metabolic rate, radiant temperature, humidity and air speed. IAQ and thermal comfort are the two important features of indoor environment of a building that has significant consideration by current building designers. International and local standards recommend conditions envisioned to adoptive situations that are satisfactory to residents [17–19]. Majority of the people spend an average of 87% of their time within enclosed environments. It has been acknowledged that the inhalation of indoor air is the major determinant of human exposure to numerous pollutants [20]. Providing improved thermal comfort settings must go simultaneously with the disparagingly vital better goal of plummeting the level of energy use in the green buildings [21]. The conventional buildings could have huge source of contamination, but the situation differs in green building; health, aesthetics and economics have become more imperative for good ventilation and better thermal comfort [22]. Thermal atmospheres, satisfaction and suitability are all prejudiced by the contest among one's prospects regard to the indoor environment in a specific situation [23]. Excessive energy savings attained by permitting air-conditioning process a broader variety of the indoor temperature variation [24]. There have been numerous factors affecting thermal comforts such as behavioural and cultural aspects, space layout, option of regulate over the situation, thermal antiquity and separate predilections [25].

2 Methodology

2.1 Measurement of Indoor Air Quality

Through three ways, IAQ could be improved, i.e. ventilation, measurement of indoor air and control of source emission. Measurement of indoor air could be carried out place in new construct buildings earlier or later indoor habitation, depending upon the certification requires [4]. Liang et al. (2014) performed an environmental monitoring study in few buildings climatic change as well as to determine change in IAQ [26]. The study also included air temperature, globe temperature, relative humidity, sound level, airspeed, VOCs and concentration of CO_2, Similar study conducted by Pei et al. (2015), by using digital instruments to measure the indoor and outdoor pollutant. In his study, a questionnaire was developed to take update on resident's habit and their satisfaction [27]. Only in a few researches, the real assessment of IAQ was carried out by recording the CO_2 level or additional pollutants, and determining the rate of ventilation, so that the definite exposures in the green constructions could be defined [2, 26, 28]. In many studies, endeavouring to match green buildings and conventional buildings to regulate for perplexing issues could be sparse [29]. In China, to examine the IAQ and inhabitant gratification of green buildings, the study group had taken a dozen methodical researches of office buildings. Contineous/periodic evaluation of indoor environmental quality must be ensured through building users questionnaire survey [30].

2.2 Measurement of Thermal Comfort

Abdallah et al. (2014) carried out a study of the consequence of a combined system, i.e. windcatcher for evaporative cooling on the ventilation and IAQ for a warm and dry climate in Egypt. The aim of the study was to achieve the high performance and the compact design by means of a satisfactory range of thermal comfort for indoor environment and IAQ, and important dimension limits such as air gap, inclination angle, chimney width along with this windcatcher's width and depth were enhanced with numerical imitation [31]. According to ASHRAE standard 55 describes the amount and the rate of ventilation necessary to achieve acceptable IAQ. The natural ventilation system not only used for providing an adequate fresh air rate within the indoor environments, but also used for refining thermal comfort conditions. A numerical research of a windcatcher was conducted in semiarid and arid zone of Mexico for the thermal comfort for residential buildings [32]. In order to identify the thermal comfort environments, the mean temperature and velocity of various zones inside the residential building were measured [33, 34].

Heat balance models (HBM): It is considered that the thermoregulatory system of human body must keep the fundamentally continuous inside temperature of the body. Analogously, the consequences of the instant thermal comfort environment could be arbitrated by the mass and heat transmission among the bodies and the neighbouring environment. To sustain a continual body temperature, people should reply physiologically to the thermal unevenness with its environment for thermal comfort inside the building. It was presumed that thermal sensations for inhabitant, i.e. feeling cold and hot were normally comparative to the extent of these answers must measure the mean skin temperature and latent heat loss which must come in the standard temperature and relative humidity range. The IAQ guidelines have been set and describe thermal comfort parameters and the control of known and specifiable contaminants to achieve acceptable IAQ [35, 36].

2.3 Problems Associated with Green Building

There are five major issues which affect the indoor environment of the green building such as inter-individual variances and satisfaction, climate situation, the role of nations, outside thermal objectivity and inadequate thermal comfort [23, 35]. It was found that resident with "pro-environmental" defiance towards more "forgiving" to accept their instantaneous environment inside the green buildings [37]. Comforting culturally persuaded clothing standard and inhabitant prospects of meticulously measured the indoor environments to substantial development in attaining an appropriate equilibrium between the energy use, thermal comfort and minimum impact on the environment [38]. Additional effort in the cultural and socio-economic area must essential, and post-tenancy assessment of current environment and the equivalent energy ingesting to make better thoughtful regarding the issues influencing indoor environment [35]. Issues arise like sick building syndrome (SBS), illness and increase in indoor pollutants could affect the overall efficiency of the inhabitants. Various illnesses which are related to mental health and the illnesses which are not simply perceptible in the limited terms could cause major difficulties in the lengthy term, i.e. cardiovascular diseases, obesity and asthma due to poor IAQ [39, 40]. Therefore, in extremely polluted cities, old people and kids/infants could be affected due to microclimatic circumstances considered as thermal discomfort which in the body for longer time intervals. Few research studies also carriedout to resolve the different natural ventilation systems by comparing the thermal comfort indices within the buildings [41]. According to ASHRAE Standard 62.1 (2013), describes the practice of ethics in building through codes and guidelines essential ventilation rates, filtration, outside air quality and the precise pollutants of attention and pollutant concentration parameters [16, 42]. The main problem for energy and thermal comfort regulation in building mechanization was to stability the struggle between the people's comfort and the energy consumption. Therefore, the thermal discomfort was happened due to the low relative humidity [43, 44].

2.4 Solutions for Thermal Comfort and Indoor Air Quality

The energy efficiency of the green building measured the reduction in the cooling and heating loads by enhancing envelope's thermal integrity. It also raises the effectiveness of cooling and heating of the equipment and depleting system's energy through efficient control methods [42]. Solution for newly constructed buildings such as removal of specific source emissions from the building materials, air cleaning through chemical and physical process and by increasing natural or mechanical ventilation systems [13]. Selection of furnishings and other building materials must have low toxicity. The control approaches must have implemented to decrease the amount of pollutant in an indoor environment below the standard limits. Basically, there were three strategies utilized for improving the IAQ inside the green buildings such as source elimination, weakening of the interior pollutants by ventilation and control of local source (Table 1).

3 Conclusion

IAQ is the major issue facing by the occupants, and it may cause due to daily activity and its specific source emission within the building. The concentrations of indoor air pollutants mainly depend upon their generation rate, volume of the indoor environment, mixing efficiency in the indoor space and the decay rates of the pollutants and improper ventilation and inadequate design of the buildings. This study suggested some strategies by providing a proper natural or mechanical ventilation system for both the IAQ and thermal comfort through which these problems can be solved.

Table 1 Strategies used for improving the IAQ [42]

Approaches	Discussion
Ventilation through heat recovery	The rate of outdoor air is maintained
Outdoor air systems	Reduce energy consumption and enhance IAQ with simplifying controls. More suppleness in cooling and heating strategies
Ventilation through displacement	Outdoor air is less, improved IAQ in living zone but not relevant in entire spaces
Natural ventilation	Mechanical cooling is less, outdoor air is more, improve thermal comfort and IAQ
Envelope tension	Permeation is depraved for IAQ and thermal comfort
Adequate distribution of air	Energy efficiency, thermal comfort and good IAQ could be achieved
More efficient particle filtration	Improved equipment efficiency, cleaner supply air filter installation and maintenance critical
Source control and make the lower ventilation	Outdoor air is less with the same or improved IAQ and thermal comfort Source characterization must necessary

Thus, the reduction of air pollutant such as NO_X, SO_X, PM_{10}, VOCs, CO, CO_2 can be observed. The study provides the detail insights of the health issues caused due to poor IAQ which may be acute or chronic in nature. Overall, a green building can attain the full efficiency with respect to IAQ and thermal comfort by considering the rule and guidelines given the various standard codes of international or regional levels.

References

Council UGB (2009) Green building design and construction: LEED reference guide for green building design and construction. US Green Building Council

Xiong Y, Krogmann U, Mainelis G, Rodenburg LA, Andrews CJ (2015) Indoor air quality in green buildings: a case-study in a residential high-rise building in the northeastern United States. J Environ Sci Health Part A 50(3):225–242. https://doi.org/10.1080/10934529.2015.981101

ASHRAE Handbook AF (2009) American society of heating, refrigerating and air-conditioning engineers. Inc., Atlanta, GA, USA

Wei W, Ramalho O, Mandin C (2015) Indoor air quality requirements in green building certifications. Build Environ 92:10–19. https://doi.org/10.1016/j.buildenv.2015.03.035

He C, Morawska L, Hitchins J, Gilbert D (2004) Contribution from indoor sources to particle number and mass concentrations in residential houses. Atmos Environ 38(21):3405–3415. https://doi.org/10.1016/j.atmosenv.2004.03.027

Fabian MP, Stout NK, Adamkiewicz G, Geggel A, Ren C, Sandel M, Levy JI (2012) The effects of indoor environmental exposures on pediatric asthma: a discrete event simulation model. Environ Health 11(1):1–15. https://doi.org/10.1186/1476-069X-11-66

Colton MD, MacNaughton P, Vallarino J, Kane J, Bennett-Fripp M, Spengler JD, Adamkiewicz G (2014) Indoor air quality in green vs conventional multifamily low-income housing. Environ Sci Technol 48(14):7833–7841. https://doi.org/10.1021/es501489u

OSG (Office of the Surgeon General) US (2005) Report of the surgeon general's workshop on healthy indoor environment: January 12–13, 2005. National Institutes of Health, Bethesda, MD

WHO (2010) WHO guidelines for indoor air quality: selected pollutants. WHO Regional Office for Europe, pp 1–484

Persily AK, Emmerich SJ (2012) Indoor air quality in sustainable, energy efficient buildings. Hvac&R Res 18(1–2):4–20. https://doi.org/10.1080/10789669.2011.592106

ASHRAE A (2009) IAQ guide: best practices for design, construction, and commissioning

Kim SS, Kang DH, Choi DH, Yeo MS, Kim KW (2008) Comparison of strategies to improve indoor air quality at the pre-occupancy stage in new apartment buildings. Build Environ 43(3):320–328. https://doi.org/10.1016/j.buildenv.2006.03.026

Heerwagen J (2000) Green buildings, organizational success and occupant productivity. Build Res Inf 28(5–6):353–367. https://doi.org/10.1080/096132100418500

Tharim AHA, Samad MHA (2016) Assessment framework of building facade in optimizing indoor thermal comfort of green building index (GBI) certified office building. In: MATEC web of conferences, EDP Sciences, pp 1–6. https://doi.org/10.1051/matecconf/20164704001

De Gracia A, Barreneche C, Fernández AI, Cabeza LF (2016) The state of the art for technologies used to decrease demand in buildings: thermal energy storage. In: Energy performance of buildings. Springer, Cham, pp 319–348. https://doi.org/10.1007/978-3-319-20831-2_16

ASHRAE Standard 55 (2013) Thermal environmental conditions for human occupancy. American Society of Heating, Refrigerating and Air-Conditioning Engineers, Inc. ISSN 1041-2336

De Dear RJ (1998) A global database of thermal comfort field experiments. ASHRAE Trans 104:1141

Fisk WJ (2000) Health and productivity gains from better indoor environments and their relationship with building energy efficiency. Annu Rev Energy Env 25(1):537–566

Huizenga C, Abbaszadeh S, Zagreus L, Arens EA (2006) Air quality and thermal comfort in office buildings: results of a large indoor environmental quality survey, pp 393–397

Roberts T (2016) We spend 90% of our time indoors. Building green. Retrieved from https://www.buildinggreen.com/blog/we-spend-90-our-time-indoors-says-who. Accessed on 27 Apr 2019

Brager G, Zhang H, Arens E (2015) Evolving opportunities for providing thermal comfort. Build Res Inf 43(3):274–287. https://doi.org/10.1080/09613218.2015.993536

Spengler JD, Chen Q (2000) Indoor air quality factors in designing a healthy building. Annu Rev Energy Env 25(1):567–600

De Dear RJ, Brager GS (2002) Thermal comfort in naturally ventilated buildings: revisions to ASHRAE Standard 55. Energy Build 34(6):549–561

Hoyt T, Lee KH, Zhang H, Arens E, Webster T (2005). Energy savings from extended air temperature setpoints and reductions in room air mixing, pp 1–5

Rupp RF, Vásquez NG, Lamberts R (2015) A review of human thermal comfort in the built environment. Energy Build 105:178–205. https://doi.org/10.1016/j.enbuild.2015.07.047

Liang HH, Chen CP, Hwang RL, Shih WM, Lo SC, Liao HY (2014) Satisfaction of occupants toward indoor environment quality of certified green office buildings in Taiwan. Build Environ 72:232–242. https://doi.org/10.1016/j.buildenv.2013.11.007

Pei Z, Lin B, Liu Y, Zhu Y (2015) Comparative study on the indoor environment quality of green office buildings in China with a long-term field measurement and investigation. Build Environ 84:80–88. https://doi.org/10.1016/j.buildenv.2014.10.015

Newsham GR, Birt BJ, Arsenault C, Thompson AJ, Veitch JA, Mancini S, Galasiu AD, Gover BN, Macdonald IA, Burns GJ (2013) Do 'green' buildings have better indoor environments? New evidence. Build Res Inf 41(4):415–434. https://doi.org/10.1080/09613218.2013.789951

Steinemann A, Wargocki P, Rismanchi B (2017) Ten questions concerning green buildings and indoor air quality. Build Environ 112:351–358. https://doi.org/10.1016/j.buildenv.2016.11.010

Lin B, Liu Y, Wang Z, Pei Z, Davies M (2016) Measured energy use and indoor environment quality in green office buildings in China. Energy Build 129:9–18. https://doi.org/10.1016/j.enbuild.2016.07.057

Abdallah ASH, Hiroshi Y, Goto T, Enteria N, Radwan MM, Eid MA (2014) Parametric investigation of solar chimney with new cooling tower integrated in a single room for New Assiut city, Egypt climate. Int J Energy Environ Eng 5(2–3):1–9. https://doi.org/10.1007/s40095-014-0092-6

Cruz-Salas MV, Castillo JA, Huelsz G (2014) Experimental study on natural ventilation of a room with a windward window and different windexchangers. Energy Build 84:458–465. https://doi.org/10.1016/j.enbuild.2014.08.033

Mostafaeipour A, Goudarzi H (2016) Economical evaluation of passive systems for residential buildings in Kerman, Iran, pp 1267–1279

Jomehzadeh F, Nejat P, Calautit JK, Yusof MBM, Zaki SA, Hughes BR, Yazid MNAWM (2017) A review on windcatcher for passive cooling and natural ventilation in buildings, Part 1: Indoor air quality and thermal comfort assessment. Renew Sustain Energy Rev 70:736–756. https://doi.org/10.1016/j.rser.2016.11.254

Yang L, Yan H, Lam JC (2014) Thermal comfort and building energy consumption implications–a review. Appl Energy 115:164–173. https://doi.org/10.1016/j.apenergy.2013.10.062

ASHRAE Standard 55 (2010) Thermal environmental conditions for human occupancy. American Society of Heating, Refrigerating and Air-Conditioning Engineers, Inc. (1). ISSN 1041–2336

Leaman A, Bordass B (2007) Are users more tolerant of 'green' buildings. Build Res Inf 35(6):662–673. https://doi.org/10.1080/09613210701529518

Fountain M, Brager GS, de Dear R (1996) Energy Build 24:179–182

Houtman I, de Jong TD, Meeuwsen JM, Brekelmans F (2008) New forms of physical and psychosocial health risks at work. Policy Department Economic and Scientific Policy, PE 408.569. Retrieved from http://www.europarl.europa.eu/document/activities/cont/201107/20110718ATT24294/20110718ATT24294EN.pdf

Jaakkola MS, Quansah R, Hugg TT, Heikkinen SA, Jaakkola JJ (2013) Association of indoor dampness and molds with rhinitis risk: a systematic review and meta-analysis. J Allergy Clin Immunol 132(5):1099–1110. https://doi.org/10.1016/j.jaci.2013.07.028

Golasi I, Salata F, de Lieto Vollaro E, Coppi M, de Lieto Vollaro A (2016) Thermal perception in the mediterranean area: comparing the mediterranean outdoor comfort index (moci) to other outdoor thermal comfort indices. Energies 9(7):550. https://doi.org/10.3390/en9070550

Persily A (2015) Challenges in developing ventilation and indoor air quality standards: the story of ASHRAE Standard 62. Build Environ 91:61–69. https://doi.org/10.1016/j.buildenv.2015.02.026

Yang R, Wang L (2012) Multi-objective optimization for decision-making of energy and comfort management in building automation and control. Sustain Cities Soc 2(1):1–7. https://doi.org/10.1016/j.scs.2011.09.001

Amin M, Dina N (2015) The impact of indoor comfort on students' experiential learning in engineering education laboratories. Doctoral dissertation, Universiti Tun Hussein Onn Malaysia

Heterogeneous Photocatalysis for Indoor Air Purification: Recent Advances in Technology from Material to Reactor Modeling

Mohan V. Lekshmi, S. M. Shiva Nagendra and M. P. Maiya

Abstract Heterogeneous photocatalytic oxidation (PCO) has attracted much attention in indoor air-purification applications. Recently, researches focus on developing novel photocatalyst based filters for integrating with the heating, ventilation, and air conditioning (HVAC) systems as well as portable air purifiers. A comprehensive knowledge on factors influencing the indoor volatile organic compounds (VOCs) degradation has been established both in bench-scale and pilot-scale experiments. This paper reviews the current status of PCO material technologies, coating methods, performance test methods, and modeling for real-world indoor air-purification application. Due attention to the basic principle of PCO and the effect of operating parameters is provided, followed by a discussion on the modes of PCO application for buildings. The review also concentrates on the practical limitations in scaling-up PCO air purifiers for large-scale applications. Some recommendations for the future research on improving material selection and reactor design to minimize by-product generation and to promote commercialization are also discussed.

Keywords Titanium dioxide · Volatile organic compounds · PCO building material · HVAC filter · Air purifier

1 Introduction

Indoor air quality (IAQ) has attracted much research interest in the past decades. Poor IAQ has a significant influence on occupant satisfaction, comfort, productivity, and health. Previously, it was reported that the indoor levels of certain air contaminants such as the volatile organic compounds (VOCs) could be five times higher

M. V. Lekshmi · S. M. Shiva Nagendra (✉)
Environmental and Water Resources Engineering Laboratory, Department of Civil Engineering, Indian Institute of Technology Madras, Chennai 600036, India
e-mail: snagendra@iitm.ac.in

M. P. Maiya
Refrigeration and Air Conditioning Laboratory, Department of Mechanical Engineering, IIT Madras, Chennai 600036, India

A. Sharma et al. (eds.), *Indoor Environmental Quality*, Lecture Notes in Civil Engineering 60,
https://doi.org/10.1007/978-981-15-1334-3_16

than outdoor air. VOCs are major class of indoor air pollutants common in domestic and commercial buildings. In indoors, VOCs can be majorly emitted from furniture, building decorations, consumer products, cooking, smoking or outdoor traffic [1]. The adverse health implications of VOCs range from sick building syndrome (SBS) symptoms and respiratory illnesses to even cancer and death. Also, VOCs take part in secondary organic aerosol (SOA) formation in indoor air and thus, exacerbates occupant exposure to undesirable air pollutants. Several air-cleaning technologies have been studied in the past for VOC abatement, i.e., adsorption, non-thermal plasma, ozonation, ultraviolet germicidal irradiation, etc. [2–4]. While adsorption, the typically used technology for VOC removal requires regular filter replacement, advanced oxidation processes such as photocatalytic oxidation (PCO) completely mineralize organic pollutants rather than accumulating them and hence, has a longer lifetime. Other advanced technologies such as non-thermal plasma and ozonation were studied for potential VOC removal. However, the high energy consumption and toxic by-product (i.e., ozone) formation from such technologies have posed a significant challenge.

Heterogeneous photocatalysis has emerged as a promising technology for indoor air-purification applications due to its room-temperature operation, non-selectivity to a wide range of indoor VOCs, and cost-effectiveness [5]. Although numerous researches have reported on various aspects of PCO for indoor air purification, namely material development [6], effect of controlling parameters [7], modeling [8], by-product quantification [9] and scale-up [10], the commercialization and practical application of the technology are still in developmental stages.

PCO uses a semiconductor material as catalyst and a light source to convert organic pollutants into harmless end products, namely CO_2 and H_2O [5]. Recently, several novel photocatalysts are being investigated for indoor air purification [6, 11, 12]. Not to mention, titanium dioxide (TiO_2) is still the most common photocatalyst reported in studies, mainly due to its stability, easy availability, non-toxicity, and relatively less cost [13]. However, overall degradation efficiency using TiO_2 depends upon reducing the recombination rate of charge carriers (electrons and holes) and expanding photoactivity in visible light spectra [14]. Several modifications to TiO_2 photocatalyst, namely doping, coupling, and addition of co-adsorbent have been reported to overcome its wide band gap and improve photodegradation efficiency in the visible region [13].

PCO technology is investigated for building applications due to its air purification, self-cleaning, and anti-bacterial properties [15]. TiO_2-coated filter are studied for use in portable air purifiers and in-duct heating, ventilation, and air conditioning (HVAC) systems [15, 16]. The application of photocatalyst integrated building materials is not uncommon since the early 1990s. TiO_2 has been investigated for its use in construction materials (pavement blocks, concrete, and tiles) and furnishings (cement mortar, wallpapers, glass, paint, etc.) due to its mechanical stability and photocatalytic activity [15].

However, the practical application of PCO suffers from two important drawbacks, namely by-product formation and accumulation of intermediates on catalyst surface progressively leading to catalyst deactivation and reduction in degradation efficiency.

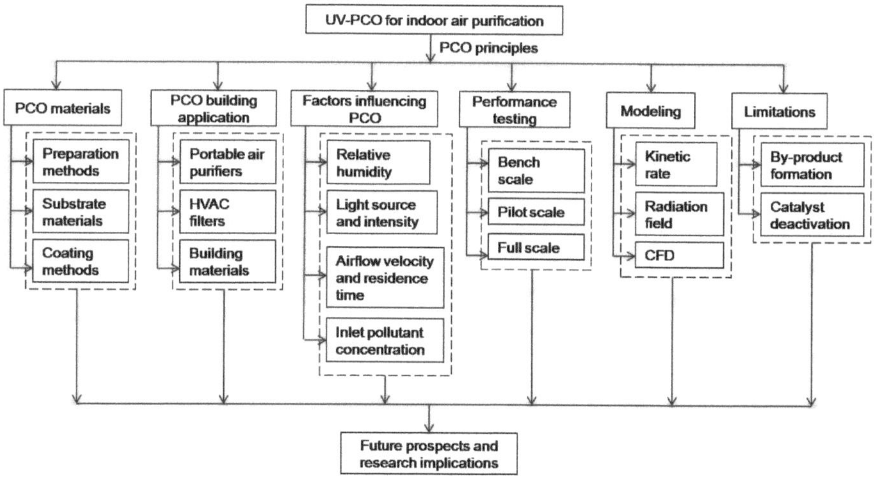

Fig. 1 Review framework of PCO in indoor air purification

The past studies have not addressed the issues in scaling-up PCO technology for real-world applications. Thus, there is a need for research in the development of an efficient PCO purifier for large-scale indoor air-purification applications.

The present paper intends to review the current advances in the area for PCO for indoor air purification including photocatalyst materials, influencing factors, building applications, test methods, and models available as shown in Fig. 1. The review, finally, comprehends the limitations and the future scope of PCO for indoor air-cleaning applications.

2 UV-PCO Principles

The photodegradation initiates when the electrons are excited by incident light energy exceeding the band gap of semiconductor material (Fig. 2). As the electron gets shifted from valence band (VB) to conduction band (CB), a hole is formed which acts as a powerful oxidizing agent. In this way, the electron–hole pairs are formed. The excited electrons reduce O_2 into superoxide radical $O_2^{\bullet-}$ and the holes (h^+) oxidizes water molecule into hydroxyl radical ($OH^{\bullet}$). The charge pairs are formed in femtoseconds (fs) and recombined within a few tens of nanoseconds [13]. The effectiveness of VOC degradation, therefore, lies on effective charge transfer from the semiconductor surface to the organic contaminants.

For TiO_2 in the anatase form, UV light ($\lambda = 360$ nm) can be suitable to excite its band gap of 3.2 eV. While excited with UV irradiation, TiO_2 forms electron–hole pairs as given in Eq. (1).

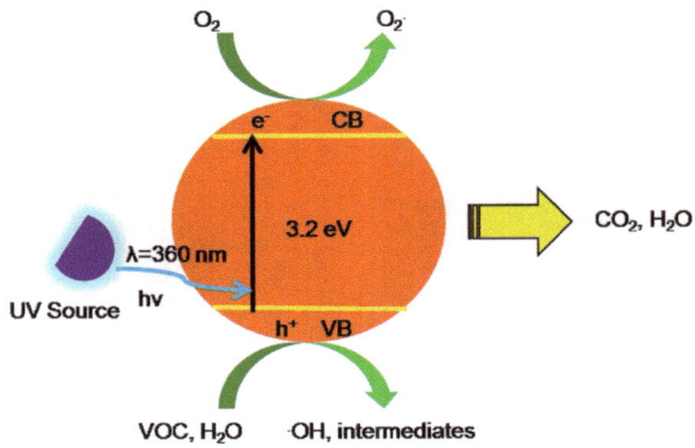

Fig. 2 Schematic representation of PCO mechanism

$$TiO_2 + h\vartheta \rightarrow h^+ + e^- \tag{1}$$

The photogenerated electrons must be accepted by electron acceptors within the few nanoseconds before it recombines with the holes to reduce charge recombination. Nanostructured materials such as nanorods, nano-fibers, nanotubes, and nanospheres can reduce the transfer distance between the surface and electron acceptors and are, hence, preferred [13]. Various parameters such as surface area, pore size, crystalline phase, and size affect the photocatalytic performance.

3 PCO Materials

Among semiconductor photocatalysts, TiO_2 has become attractive in air and water treatment due to its non-toxicity, relatively less cost, photo-stability, and rapid electron transfer to molecular oxygen [17]. TiO_2 in both pure and modified form has been proven to be effective for a wide range of indoor VOCs.

As discussed in the previous section, the photocatalytic activity of catalyst depends on the crystallinity of TiO_2 [18]. TiO_2 photocatalysts are available in three crystal phases namely, anatase, rutile, and brookite. The former type is more active than rutile due to higher electron-hole pair generation, greater affinity toward O_2, higher number of surface hydroxyl groups, and lower recombination rate [19]. In this regard, degussa-P25 TiO_2 is the most commonly used photocatalyst in past studies due to the presence of both anatase and rutile phases (75% anatase and 25% rutile), rutile particle band bending in presence of anatase phase, and single crystallites nature of P25 [20–22]. Besides the crytallinity, the surface area and porosity of TiO_2 significantly influence the photoactivity. Larger catalyst surface area and smaller pore size is known to enhance the photocatalytic activity [23]. Photocatalyst loading density on the support is another major factor influencing the decomposition of pollutants as it determines the available catalyst surface area and number of active sites. The higher the catalyst layer thickness, higher is catalyst surface area available for adsorption of

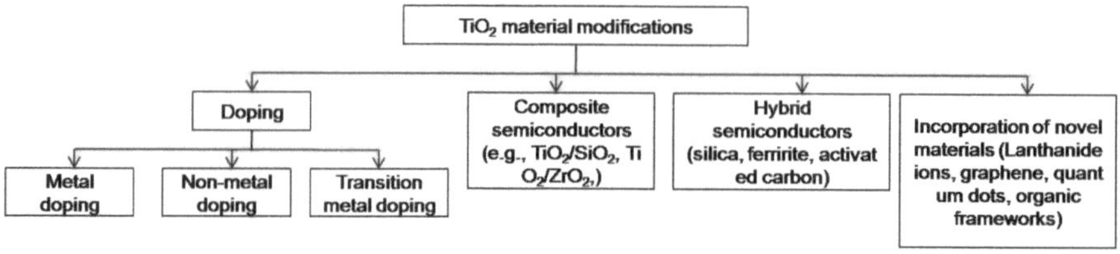

Fig. 3 PCO material modifications

contaminants [24]. However, excessive catalyst loading can cause shadowing effect and masks part of catalyst surface wherein no charge carriers are generated. Several studies have reported optimum catalyst loading beyond which the photocatalytic activity declines [23, 25, 26].

Recent studies are focused on developing visible light-responsive photocatalysts which can be easily employed indoors and inside vehicles. Doping the photocatalyst is a surface modification technique to lower the band gap energy of catalyst and shift toward the visible light region. Several studies on doping TiO_2 with metals (Mn [27], Ni [28], Fe [29], and Pt [30]) and non-metals (C [31], N [32], and F [33]) are available. The metal ion dopants in the crystalline matrix of TiO_2 significantly improve photoactivity apart from reducing recombination rate [34]. Similarly, binary catalysts are developed by integrating TiO_2 with other semiconductors such as WO_3, Ag_3VO_4, SiO_2, SnO_2, $MnCO_3$, CdS, ZnO, and porous materials such as activated carbon with the aim to expand its photo-activation into visible spectra. The different TiO_2 modifications reported in previous studies are summarized in Fig. 3.

4 Preparation Methods of the Photocatalyst

The surface properties of the semiconductor material determine its photocatalytic activity [35]. The physical properties of the catalyst such as pore size, pore thickness, and distribution and crystalline structure need to be tuned which greatly depends upon the preparation technique adopted. Various preparation methods of TiO_2 photocatalyst have been reported in published literature, namely sol–gel method [26, 27], vapor decomposition of titanium alkyloxides [36], gas-phase hydrolysis of $TiCl_4$ [37], and hydrothermal technique [38]. Yu et al. [35] prepared a porous TiO_2 coating using sol–gel method using tetrabutylorthotinate as precursor and polyethylene glycol as organic polymer followed by calcination at 520 °C. A similar study by Martins et al. [39] evaluated the photocatalytic performance of TiO_2-coated AC prepared by sol–gel technique compared to TiO_2 P25. The TiO_2-coated AC obtained using titanium isopropoxide as precursor followed by pyrolysis at 500 °C showed better structural properties for photodegradation [39]. Rubio et al. [36] prepared titania powder by gas-phase hydrolysis of titanium tetrabutroxide/butanol mixture in a pressurized (N_2 atmosphere) and thermostatically controlled glass container at 180 °C.

In the process, titanium tetrabutroxide and water were sprayed inside the hydrolysis chamber (25°) where they react to form TiO_2 particles which was further calcinated at 450 °C for 2 h [36]. In the case of hydrothermal method, the reactions are carried out in a closed vessel containing the aqueous solution and the products are recovered and used after attaining room temperature. Liu et al. [38] prepared a highly active TiO_2/AC composite by adding $TiCl_4$ dropwise into $(NH_4)_2SO_4$/HCl/water mixture followed by heating at 98 °C for 1 h. Among various techniques, the sol–gel method has been recognized as an efficient and frequently reported method to fabricate a porous coating of TiO_2 [35].

5 PCO Substrates and Coating Techniques

Catalyst immobilization onto the substrate is one of the key determinants affecting PCO efficiency for degrading indoor air pollutants. The coating technique and type of substrate need to be carefully selected for effective immobilization. In previous studies, substrates such as carbon nanotube [40], activated carbon filter (ACF) fibers [41], zeolites [42], glass [43], glass fiber filter (FGF) [44], stainless steel [45], silica materials [46] etc. have been used as substrate. Activated carbon filters have high porosity with higher surface area for adsorption and enhance the photocatalytic activity against organic pollutants [47, 48]. Some commonly used carbon substrates in previous studies are carbon felt (ACFF) [49, 50], carbon cloth [44], carbon paper [51], carbon powder [45], granular carbon [52], graphene [53] etc. Studies have compared the photocatalytic activity of TiO_2/ACF against TiO_2/FGF systems and concluded that the synergetic adsorption–photodegradation using the former has a higher efficiency for gas-phase VOCs removal [10, 12, 47].

In the case of PCO building materials, a TiO_2-coating on the glass, concrete blocks, filter, etc., was obtained by dispersing the nanoparticles solvents such as ethanol, polymer matrix, etc., followed by ultrasonication treatment to form a uniform suspension. The PCO-coating technique adopted should provide a stable coating, proper contact between the catalyst and pollutant molecules, be cost-effective and non-selective to substrate types. Commonly adopted coating techniques are dip coating [54] and spray coating [55]. Other methods such as electrophoretic deposition (EPD) [56], sol–gel method [57], and chemical vapor deposition (CVD) [58] are also reported in published literature. The dip-coating technology is an especially suitable method for flat substrates (concrete blocks, glass, etc.) to form a uniform thin layer of TiO_2 from the suspension [59] (Table 1).

6 PCO for Indoor Air Purification

Photocatalytic oxidation is an economical and effective method for degrading organic pollutants and bioaerosols in indoor air. The technology is characterized by lower

Table 1 Effect of substrate material on removal efficiency of VOCs

Pollutant, concentration	Photocatalyst/substrate	Removal efficiency (%)		Reference
		With substrate	Pure TiO_2	
BTEX, 20 ppb	Activated carbon felt	Benzene: 74; Toluene: 78; Ethylbenzene: 73; Xylene: 86	Benzene: 57; Toluene: 69; Ethylbenzene: 67; Xylene: 72	[41]
Formaldehyde, 1 ppm	Activated carbon filter	79.4	25.7	[50]
Acetaldehyde and toluene, 5 ppm	Activated carbon powder	Acetaldehyde: 60, Toluene: 47	Acetaldehyde: 75, Toluene: 36	[43]
Toluene, 230–1150 ppm	Activated carbon felt	57–100	41	[47]
Toluene, 80–174 ppm	Activated carbon felt	100	65.4	[60]

power consumption and maintenance requirements [34]. However, high recombination rate of charge carriers (electrons and holes) and limited photocatalytic activity under visible light is the major drawback of the process and reduces the overall degradation efficiency [13]. A few studies reported TiO_2-coated activated carbon filters for HVAC applications which showed enhanced photocatalytic activity due to the large surface area of the catalyst and lower charge recombination [47, 46, 60, 61]. However, past studies were conducted in laboratory-scale reactors and scaling up of technology to operate under real-world conditions has many limitations. For example, most studies considered low airflow velocities and high inlet pollutant concentration (ppm) which is very different from realistic conditions [62]. Therefore practical design and application of PCO for indoor air purification need more extensive research addressing these knowledge gaps.

7 Portable Air Purifiers

An efficient PCO air purifier should have a low-pressure drop, sufficient mass transfer between the gas phase to the solid catalyst, and better light utilization [16]. Therefore, a PCO air purifier uses an air mover, catalyst supported on the suitable substrate and a light source. Past studies investigated the effectiveness of PCO air purifiers to remove individual VOC and mixture of VOCs present in real-world indoor conditions. Kolarik et al. [63] evaluated the improvement in perceived indoor air quality using a PCO air purifier in test rooms doped with typical building chemicals. The study showed that the use of PCO air purifier significantly improved the perceived air quality, reduced odor, and pollutant levels. Another study by Gunschera et al.

[64] tested PCO air cleaners in 24 and 48 m^3 test chambers for degradation of sub-ppm levels of common indoor VOCs (aldehyde, toluene, α-pinene, decane, etc.). The study concluded that significant levels of by-products (aldehyde and acetone) formed during reaction need to be addressed carefully before real-world application. However, portable PCO purifiers can be easily deployed and maintained when compared with PCO integrated into HVAC ducts.

8 PCO Coated HVAC Filter

Urban buildings are provided with HVAC systems to provide ventilation and maintain adequate IAQ. For this purpose, HVAC ducts are equipped with various air-cleaning technologies operating at lower energy costs. An extensive research database on the application of PCO for indoor air-cleaning in an HVAC system based on laboratory-scale experiments has been published to date [22–24]. The study results showed that PCO is attractive for integrating with existing and new HVAC systems due to its room-temperature operation, modular design, lower pressure drop, and better mass transfer rate [18, 25]. Yu et al. [65] tested the VOC removal efficiency of PCO filter under various conditions (air change rate = 0.5–1.5 h^{-1} and RH = 30–70%) and reported that efficiency increased with RH and decreased with filter face velocity. Yang et al. [16] developed TiO_2-coated foam nickel air purifier for HVAC systems and optimized the design and energy consumption. The study pointed out that adjusting the purifier configuration can increase the residence time of pollutants and decrease the flow resistance. Therefore, there is a need to develop an efficient PCO technology for integrating with HVAC ducts and reducing building energy consumption. However, a knowledge gap exists on the behavior of the PCO air cleaners under real-world conditions such as high air flow rates, low residence time, ppb level pollutant concentration, and presence of a mixture of organic vapors.

9 Building Materials

TiO_2 photocatalysts can be integrated with building materials such as concrete without altering its properties and can be activated under solar irradiation. It can be used in other forms such as cement mortar [66], glass [67], wallpapers [68], paint [69], etc. Photocatalytic paints are the most commonly used PCO material for improving indoor air quality. Maggos et al. [70] tested a TiO_2-containing paint for degrading NO_x gas emitted in a closed parking area whose ceiling was covered with acrylic TiO_2-containing paint. The results indicated a significant reduction in NO (19%) and NO_2 (20%) levels. Auvinen and Wirtanen [69] investigated six different photocatalytic paints containing different binders such as lime, silica sol–gel, polyorganic siloxane, and organic binders for degrading formaldehyde as well as a mixture of indoor VOCs. The study did not show any significant removal of VOCs but showed an increase in formaldehyde levels due to the reaction. Also, the study also examined different substrates such as glass, polymeric plaster, and gypsum which did not have any influence on photocatalytic activity.

Similarly, another study by Kolarik and Toftum [71] examined improvement in perceived air quality using a cement-based photocatalytic paint coated on gypsum board (13 m^2). The study could not conclude any improvement in perceived air quality using photocatalytic paint. Also, the photocatalytic activity of TiO_2 can reduce the long-term durability of polymeric paints. Recent studies reported that photocatalytic materials such as lanthanum and graphene-doped TiO_2 could be used for paints which can suppress and tune photoactivity of TiO_2 to balance between the mechanical stability of coating and improving indoor air quality [29, 30].

10 Factors Influencing PCO

This section briefly explains the effects of relative humidity, airflow velocity, light source, and pollutant characteristics on the photodegradation of gas-phase VOCs. The comprehensive review of the factors affecting PCO is available in Mamaghani et al. [72].

11 Relative Humidity (RH)

RH can have either a positive or negative influence on PCO efficiency. In other words, the water molecules act both as a precursor for the formation of hydroxyl radical (oxidizing agent) and as a competitor for active adsorption sites on TiO_2 surface. The competitive adsorption reduces the adsorption capacity of VOC and reduces the reaction rate. Therefore, change in RH can have a significant impact on PCO efficiency. Some studies have reported improvement in PCO efficiency for toluene and benzene oxidation with an increase in RH [73]. A similar effect on benzene removal was achieved by Bouazza et al. [74]. However, the propene conversion efficiency reduced with an increase in RH. Another study by Mo et al. [75] reported a negative effect of an increase in RH on toluene removal efficiency with inlet concentration in the range of 90–250 ppb. However, at higher toluene concentrations (400–800 ppb), an optimum RH value was reached achieving maximum removal efficiency [75]. Additionally, doped TiO_2 can improve the photocatalytic efficiency when compared to undoped TiO_2 at the same RH level [13]. Therefore, it can be concluded that the effect of RH on conversion efficiency depends upon the type and level of pollutant and also the type of photocatalyst used.

12 Airflow Velocity and Residence Time

Airflow rate is a key determinant affecting the PCO efficiency for VOC oxidation. With the increase in airflow velocity, two antagonistic effects were brought forth, namely the decrease in residence time of the pollutant molecules near catalyst surface and the increase in the mass transfer of pollutant molecules from bulk to catalyst surface [76]. The reduction in residence time inside the photoreactor reduces the adsorption of pollutant molecules and lowers the photodegradation efficiency whereas the increase in mass transfer enhances the PCO reaction rate. But considering the airflow velocities consistent with HVAC systems, the improvement in PCO efficiency due to increase in mass transfer is offset by the significant reduction in residence time [72]. Similarly, under the shorter residence time encountered in portable air cleaners, the probability of contaminants contacting the hydroxyl radicals may be meager. Sleiman et al. [77] observed around 30% reduction in toluene removal efficiency with increase in airflow rate from 0.07 to 0.35 $m^3\ min^{-1}$. Use of the synergetic adsorption–photodegradation effect of TiO_2 combined with activated carbon can have more significant in such cases.

13 Light Source and Intensity

Light intensity is another crucial factor affecting the VOC photodegradation rate (Table 2). Generally, it is agreed that PCO efficiency increases with an increase in light intensity [78]. The increase in light intensity enhances the formation of oxidant species, enhances the photoactivity, and degradation of pollutants [79, 80]. Some

Table 2 Different light sources, their merits and demerits

Light source	Merits	Demerits	Reference
Visible light	Can degrade pollutants at visible range, ease of commercialization, lower operational cost	Lower degradation efficiency compared to UV-PCO	[82]
LED light	Inexpensive, high current to light conversion with negligible heating, longer life	Wavelength of LED is above 360 nm and therefore requires a special LED responsive catalyst	[83]
UV-PCO	Moderate to high VOC destruction efficiency	Catalyst deactivation or poisoning, needs catalyst replacement, catalyst activity only in UV region restricting use in indoors and inside vehicles	[84]
VUV PCO	Higher degradation efficiency, lower catalyst deactivation	Produces ozone as by-product	[79]

studies reported the use of vacuum UV (VUV) lamps wherein pollutant degradation takes place through photolysis, radical oxidation, and ozonation [36, 37]. The use of VUV lamps has been reported to improve the degradation efficiency compared to conventional UV lamps irradiating at 254–365 nm [7]. It was also reported that VUV lamps provide strong oxidation and prevent the formation and accumulation of intermediates which in turn reduces the catalyst deactivation [81]. However, PCO using VUV lamps produces ozone as a by-product which needs to be treated properly [36, 39].

14 Inlet Pollutant Concentration

One of the main gap in the published literature on PCO of gas-phase VOCs is that most of the past studies focused on high inlet pollutant concentrations (ppm level) whereas, in real-world indoor conditions, VOCs are present only in ppb levels. Interpolating the results from previous studies can be misleading as the rate of reaction in the presence of high inlet concentrations will be entirely different compared to low concentrations. The past studies have reported that PCO efficiency decreases with an increase in inlet pollutant concentration irrespective of the type of target VOC [31, 38]. In general, at higher pollutant concentrations, the ratio of the sum of reactive species and active sites on catalyst surface to pollutant molecules decrease resulting in incomplete degradation of VOCs.

15 PCO Performance Testing

Performance evaluation of PCO for VOC degradation under real indoor conditions needs to be carried out before using the technology for envisaged applications. Extensive research in the past focused on VOC removal using PCO in bench-scale, pilot-scale, and full-scale experiments. Most of the studies applied formaldehyde [65], toluene [47], benzene [85], and trichloroethylene [86] as typical indoor VOCs for performance evaluation. However, so far, there is no standard test method available for assessing the air-cleaning performance of photocatalytic systems. Therefore, it is difficult, the compare the result across available studies which are conducted under different environmental conditions such as RH, airflow rate, light source, etc. [87]. A recent study [10] evaluated the photocatalytic degradation of toluene and isobutanol obtained by using a bench-scale, pilot-scale, and full-scale reactor under real-world conditions (low pollutant concentration, short residence time, and high airflow rates) by maintaining the RH value at $50 \pm 5\%$. The study also compared the effect of UV light type (UVC and vacuum UV lamps) on photodegradation efficiency. The study concluded that the removal efficiency obtained using bench-scale reactors are higher than the pilot and full-scale reactors. Thus, scale-up of large-scale PCO systems for indoor air purification based on laboratory-scale results may be inaccurate. The removal efficiency also depends upon the type of reactor design adopted due to the variation in incident radiation flux as well as the contact between photocatalyst and target pollutant [59].

Clean air delivery rate (CADR) is one method typically adopted for evaluating the air-cleaning performance of PCO reactors [34]. Higher the CADR value, faster the air purifier cleans the air. For PCO air purifiers, CADR can be written as [88]:

$$\mathrm{CADR} = G \times \eta \tag{2}$$

where G is the airflow rate of the PCO air purifier, and η is the fractional VOC removal efficiency. However, this index does not account for the by-product formation from PCO chemical reactions. Therefore, extensive studies have to be conducted to investigate the absence of carcinogenic by-product (i.e., formaldehyde, benzene) formation from PCO purifiers before large-scale application.

16 PCO Modeling

An effective PCO reactor design is necessary before scale-up and commercialization of the technology [89]. A properly designed PCO system should have the following characteristics, namely low pressure drop, adequate mass transfer from the gas bulk to solid catalyst surface, proper utilization of incident radiation, low power consumption, and complete mineralization of the contaminants. Thus, design of a proper PCO reactor requires insights on the reaction kinetics (adsorption, desorption, and photodegradation), mass transfer rates from the gas phase to catalyst surface and also the photon transfer. Optimization of radiation (light intensity) and reactor design enabling effective mass transfer are inevitable for developing an energy-efficient PCO air-purification system. PCO models can be a useful tool for design, optimization as well as scale-up of PCO reactors for large-scale applications. In the past, studies have focused on various aspects of modeling, namely flow dynamics [90], reaction kinetics [76, 91], and radiation modeling [92, 93]. Recently, CFD models have been reported to accurately predict kinetic parameters along with fluid flow simulation [94, 95]. A combined modeling of radiation field along with these CFD models is required for proper design of PCO system for the future scale-up.

17 Kinetic Rate Models

The kinetic rate models give information on how the reactants are converted into final end products. PCO mechanism can be explained to be surface catalysis where reaction rate depends upon the surface coverage of pollutant molecules on the catalyst surface. The commonly used surface catalysis model is Langmuir–Hinshelwood model. The reaction rate (r) according to the L-H model can be given as follows:

$$r = \frac{K_p K_a C_o}{1 + K_a C_o} \quad (3)$$

where K_a is Langmuir adsorption constant or reaction kinetic constant in $m^3\ g^{-1}$, K_p is degradation rate constant mg $m^{-3}\ min^{-1}$, and C_o is the inlet pollutant concentration in mg m^{-3}. The value of the kinetic rate can be estimated by using curve-fitting methods. Therefore, the reaction rate models can be used to predict the speed at which pollutants can be removed by knowing the inlet pollutant concentration.

18 Radiation Field Models

The radiation field models can be used to optimize the distribution of light radiation intensity inside the PCO reactor to get uniform irradiation on the catalyst surface. For this purpose, steady-state mathematical-based models such as first-principles and predictive engineering models have been used in the previous studies [60, 51]. In general, these models combine the energy conservation and momentum conservation principles of fluid flow and pollutant species accounting for target pollutant, by-product and product concentration fields. Raupp et al. [96] reviewed the use of a radiation field model for the design of monolith PCO purifiers. Hossain et al. [90] developed a three-dimensional convection–diffusion model for monolith PCO reactors using first-principle radiation field sub model. Imoberdorf et al. [92] proposed a Monte Carlo based model coupling with the ray tracking method. The model takes into account the interaction between the TiO_2 particle and photons with due consideration toward the shadowing effect of the particles, reflection and absorption by the TiO_2 films. The study concluded that radiation field models are capable of accurately predicting the photodegradation of formic acid on catalyst surface [92]. Thus, radiation models can be used to optimize the design of air-purification systems and to scale-up for commercial applications.

19 CFD-Based Models

Previous studies have reported that CFD models can be employed to predict the PCO reaction mechanism. In addition to steady-state reaction data, the CFD-based models can provide spatial and temporal variation in pollutant removal both in bulk and catalyst surface. Verbruggen et al. [94] presented a CFD-based model for modeling adsorption and degradation of acetaldehyde on a photocatalytic filter fiber. The study concluded that the adsorption constant and maximum adsorption capacity data obtained from CFD modeling could be valuable inputs to kinetic rate models and hence, can be used for the future air purifier designs. A recent study by Roegiers et al. [95] reported a combined computational fluid dynamics (CFD)-radiation field-reaction kinetic model for modeling acetaldehyde degradation in multi-tube PCO reactor. By integrating the CFD models with L-H kinetics and uniform irradiance distribution, it will be possible to precisely model PCO air-purification systems [95].

20 Limitations

Although the use of PCO for indoor air purification has been proposed in earlier researches, there are some technical aspects that are still unanswered. The major drawbacks of PCO technology are the deactivation of the photocatalyst, low photocatalytic activity, and generation of by-products [87]. Undesirable by-products can be more toxic and irritating than the target VOCs [59]. The photocatalyst lifetime is important in overall process cost as it determines the frequency of catalyst replacement [97]. Previous studies clearly stated that PCO reactions produce by-products such as aldehydes, ketones, and organic acids besides CO_2 and H_2O. The formation of intermediate compounds and their accumulation on catalyst surface causes catalyst deactivation. This deactivation may be addressed by expanding the adsorption band of the photocatalyst to visible range, reducing the electron-hole pair recombination, and proper reactor design maximizing light utilization [87].

21 Summary and Future Perspectives

In developed and developing countries, people spend a major portion of their time in indoor environments. Therefore, providing a safe indoor space has become a priority to building designers and policy makers. Reduction in ventilation rate in modern buildings has exacerbated human exposure to harmful volatile organic compounds (VOCs). Available studies showed that VOCs are the major group of indoor air pollutants present in common indoor environments such as houses, offices, cars, and eateries. Ultraviolet photocatalysis has been reported to have immense potential for indoor air-purification applications. Despite numerous researches have focused on the technology, commercialization and scaling-up of PCO are still in its nascent stages. The knowledge gap between available laboratory-based studies and actual application of PCO technology for air decontamination needs to be addressed. Some limitations of the technology, namely by-product generation and catalyst deactivation have impeded the commercialization of technology. Certain catalyst modifications such as doping and addition of adsorbents were shown to enhance the photocatalytic activity, increase the lifetime of PCO reactors reduce deactivation, and reduce intermediate formation. Therefore, studies in the future should focus on developing novel, efficient, and environmentally friendly photocatalytic materials and reactor designs minimizing the generation of toxicants. Further, comprehensive modeling study which couples, flow dynamics, reaction kinetics, and radiation modeling is inevitable for a proper design of PCO reactor for real-world air-purification applications. The performance test methods should be standardized in the future for ease in scaling-up the PCO reactors for real-world applications.

References

1. Yu BF, Hu ZB, Liu M, Yang HL, Kong QX, Liu YH (2009) Review of research on air-conditioning systems and indoor air quality control for human health. Int J Refrig 32:3–20. https://doi.org/10.1016/j.ijrefrig.2008.05.004
2. Davidson JH, McKinney PJ (1998) Chemical vapor deposition in the corona discharge of electrostatic air cleaners. Aerosol Sci Technol 29:102–110. https://doi.org/10.1080/02786829808965555
3. Harper M (2000) Sorbent trapping of volatile organic compounds from air. J Chromatogr A 885:129–151. https://doi.org/10.1016/S0021-9673(00)00363-0
4. Waring MS, Siegel JA (2011) The effect of an ion generator on indoor air quality in a residential room. Indoor Air 21:267–276. https://doi.org/10.1111/j.1600-0668.2010.00696.x
5. Sidheswaran MA, Destaillats H, Sullivan DP, Cohn S, Fisk WJ (2012) Energy efficient indoor VOC air cleaning with activated carbon fiber (ACF) filters. Build Environ 47:357–367. https://doi.org/10.1016/j.buildenv.2011.07.002
6. Ho CC, Kang F, Chang GM, You SJ, Wang YF (2019) Application of recycled lanthanum-doped TiO_2 immobilized on commercial air filter for visible-light photocatalytic degradation of acetone and NO. Appl Surf Sci 465:31–40. https://doi.org/10.1016/j.apsusc.2018.09.136
7. Farhanian D, Haghighat F, Lee CS, Lakdawala N (2013) Impact of design parameters on the performance of ultraviolet photocatalytic oxidation air cleaner. Build Environ 66:148–157. https://doi.org/10.1016/j.buildenv.2013.04.010
8. Zhong L, Haghighat F (2018) Modeling of by-products from photocatalytic oxidation (PCO) indoor air purifiers: a case study of ethanol. Build Environ 144:427–436. https://doi.org/10.1016/j.buildenv.2018.08.048
9. Farhanian D, Haghighat F (2014) Photocatalytic oxidation air cleaner: identification and quantification of by-products. Build Environ 72:34–43. https://doi.org/10.1016/j.buildenv.2013.10.014
10. Shayegan Z, Haghighat F, Lee C (2019) Photocatalytic oxidation of volatile organic compounds for indoor environment applications: three different scaled setups. Chem Eng J 357:533–546. https://doi.org/10.1016/j.cej.2018.09.167
11. Wang S, Ang HM, Tade MO (2007) Volatile organic compounds in indoor environment and photocatalytic oxidation: state of the art. Environ Int 33:694–705. https://doi.org/10.1016/j.envint.2007.02.011
12. Coelho LL, Estrella AS, Hotza D, Li Puma G, Moreira RDFPM, de Amorim SM (2018) Modulating the photocatalytic activity of TiO_2 (P25) with lanthanum and graphene oxide. J Photochem Photobiol A Chem 372:1–10. https://doi.org/10.1016/j.jphotochem.2018.11.048
13. Shayegan Z, Lee CS, Haghighat F (2018) TiO_2 photocatalyst for removal of volatile organic compounds in gas phase—a review. Chem Eng J 334:2408–2439. https://doi.org/10.1016/j.cej.2017.09.153
14. Pichat P (2019) A brief survey of the practicality of using photocatalysis to purify the ambient air (indoors or outdoors) or air effluents. Appl Catal B Environ 245:770–776. https://doi.org/10.1016/j.apcatb.2018.12.027
15. Chen J, sun Poon C (2009) Photocatalytic construction and building materials: from fundamentals to applications. Build Environ 44:1899–1906. https://doi.org/10.1016/j.buildenv.2009.01.002

16. Yang L, Cai A, Luo C, Liu Z, Shangguan W, Xi T (2009) Performance analysis of a novel TiO_2-coated foam-nickel PCO air purifier in HVAC systems. Sep Purif Technol 68:232–237. https://doi.org/10.1016/j.seppur.2009.05.008
17. Luenloi T, Chalermsinsuwan B, Sreethawong T, Hinchiranan N (2011) Photodegradation of phenol catalyzed by TiO_2 coated on acrylic sheets: kinetics and factorial design analysis. Desalination 274:192–199. https://doi.org/10.1016/j.desal.2011.02.011
18. Li X, Yu J, Jaroniec M (2016) Hierarchical photocatalysts. Chem Soc Rev 45:2603–2636. https://doi.org/10.1039/c5cs00838g
19. Verbruggen SW (2015) TiO_2 photocatalysis for the degradation of pollutants in gas phase: from morphological design to plasmonic enhancement. J Photochem Photobiol C Photochem Rev 24:64–82. https://doi.org/10.1016/j.jphotochemrev.2015.07.001
20. Monteiro RAR, Miranda SM, Rodrigues-Silva C, Faria JL, Silva AMT, Boaventura RAR, Vilar VJP (2015) Gas phase oxidation of n-decane and PCE by photocatalysis using an annular photoreactor packed with a monolithic catalytic bed coated with P25 and PC500. Appl Catal B Environ 165:306–315. https://doi.org/10.1016/j.apcatb.2014.10.026
21. Ohno T, Sarukawa K, Tokieda K, Matsumura M (2001) Morphology of a TiO_2 photocatalyst (Degussa, P-25) consisting of anatase and rutile crystalline phases. J Catal 203:82–86. https://doi.org/10.1006/jcat.2001.3316
22. Thevenet F, Guaïtella O, Herrmann JM, Rousseau A, Guillard C (2005) Photocatalytic degradation of acetylene over various titanium dioxide-based photocatalysts. Appl Catal B Environ 61:58–68. https://doi.org/10.1016/j.apcatb.2005.03.015
23. Alonso-Tellez A, Masson R, Robert D, Keller N, Keller V (2012) Comparison of Hombikat UV100 and P25 TiO_2 performance in gas-phase photocatalytic oxidation reactions. J Photochem Photobiol A Chem 250:58–65. https://doi.org/10.1016/j.jphotochem.2012.10.008
24. Chovelon J-M, Vildozo D, Ferronato C, Sleiman M (2010) Photocatalytic treatment of indoor air: optimization of 2-propanol removal using a response surface methodology (RSM). Appl Catal B Environ J, 303–310. https://doi.org/10.1016/j.apcatb.2009.11.020
25. Choi W, Ko JY, Park H, Chung JS (2001) Investigation on TiO_2-coated optical fibers for gas-phase photocatalytic oxidation of acetone. Appl Catal B Environ 31:209–220. https://doi.org/10.1016/S0926-3373(00)00281-2
26. Monteiro RAR, Silva AMT, Ângelo JRM, Silva GV, Mendes AM, Boaventura RAR, Vilar VJP (2015) Photocatalytic oxidation of gaseous perchloroethylene over TiO_2 based paint. J Photochem Photobiol A Chem 311:41–52. https://doi.org/10.1016/j.jphotochem.2015.06.007
27. Katsanaki A, Kiriakidis G, Binas VD, Sambani K, Maggos T (2011) Synthesis and photocatalytic activity of Mn-doped TiO_2 nanostructured powders under UV and visible light. Appl Catal B Environ 113–114:79–86. https://doi.org/10.1016/j.apcatb.2011.11.021
28. Tseng HH, Wei MC, Hsiung SF, Chiou CW (2009) Degradation of xylene vapor over Ni-doped TiO_2 photocatalysts prepared by polyol-mediated synthesis. Chem Eng J 150:160–167. https://doi.org/10.1016/j.cej.2008.12.015
29. Banisharif A, Khodadadi AA, Mortazavi Y, Anaraki Firooz A, Beheshtian J, Agah S, Menbari S (2015) Highly active Fe_2O_3-doped TiO_2 photocatalyst for degradation of trichloroethylene in air under UV and visible light irradiation: experimental and computational studies. Appl Catal B Environ 165:209–221. https://doi.org/10.1016/j.apcatb.2014.10.023
30. Dong F, Wang H, Sen G, Wu Z, Lee SC (2011) Enhanced visible light photocatalytic activity of novel Pt/C-doped $TiO_2/PtCl_4$ three-component nanojunction system for degradation of toluene in air. J Hazard Mater 187:509–516. https://doi.org/10.1016/j.jhazmat.2011.01.062
31. Ren W, Ai Z, Jia F, Zhang L (2007) Low temperature preparation and visible light photocatalytic activity of mesoporous carbon-doped crystalline TiO_2. 69:138–144. https://doi.org/10.1016/j.apcatb.2006.06.015
32. Kamaei M, Rashedi H, Dastgheib S, Tasharrofi S (2018) Comparing photocatalytic degradation of gaseous ethylbenzene using N-doped and pure TiO_2 nano-catalysts coated on glass beads under both UV and visible light irradiation. Catalysts. 8:466. https://doi.org/10.3390/catal8100466

33. Li D, Haneda H, Hishita S, Ohashi N (2005) Visible-light-driven N-F-codoped TiO_2 photocatalysts. 2. Optical characterization, photocatalysis, and potential application to air purification. Chem Mater 17:2596–2602. https://doi.org/10.1021/cm049099p
34. Mo J, Zhang Y, Xu Q, Lamson JJ, Zhao R (2009) Photocatalytic purification of volatile organic compounds in indoor air: a literature review. Atmos Environ 43:2229–2246. https://doi.org/10.1016/j.atmosenv.2009.01.034
35. Yu J, Zhao X, Du J, Chen W (2000) Preparation, microstructure and photocatalytic activity of the porous TiO_2 anatase coating by sol-gel processing. https://doi.org/10.1023/a:1008703719929
36. Rubio J, Oteo JL, Villegas M, Duran P (1997) Characterization and sintering behavior of submicrometer titanium dioxide spherical particles obtained by gas-phase hydrolysis of titanium tetrabutoxide. https://doi.org/10.1023/a:1018579500691
37. Zeatoun L, Feke D (2005) Characterization of TiO_2 smoke prepared using gas-phase hydrolysis of $TiCl_4$. https://doi.org/10.1002/ppsc.200500947
38. Liu SX, Chen XY, Chen X (2007) A TiO_2/AC composite photocatalyst with high activity and easy separation prepared by a hydrothermal method. 143:257–263. https://doi.org/10.1016/j.jhazmat.2006.09.026
39. Martins AC, Cazetta AL, Pezoti O, Souza JRB, Zhang T, Pilau EJ, Asefa T, Almeida VC (2017) Sol-gel synthesis of new TiO_2/activated carbon photocatalyst and its application for degradation of tetracycline. Ceram Int 43:4411–4418. https://doi.org/10.1016/j.ceramint.2016.12.088
40. Chen J, Li G, Huang Y, Zhang H, Zhao H, An T (2012) Applied catalysis B: environmental optimization synthesis of carbon nanotubes-anatase TiO_2 composite photocatalyst by response surface methodology for photocatalytic degradation of gaseous styrene. Applied Catal B Environ 123–124:69–77. https://doi.org/10.1016/j.apcatb.2012.04.020
41. Ao CH, Lee SC (2003) Enhancement effect of TiO_2 immobilized on activated carbon filter for the photodegradation of pollutants at typical indoor air level. Appl Catal B Environ 44:191–205. https://doi.org/10.1016/S0926-3373(03)00054-7
42. Noorjahan M, Kumari VD, Subrahmanyam M, Boule P (2004) A novel and efficient photocatalyst: TiO_2-HZSM-5 combinate thin film. Appl Catal B Environ 47:209–213. https://doi.org/10.1016/j.apcatb.2003.08.004
43. Krýsa J, Baudys M, Vislocka X, Neumann-Spallart M (2018) Composite photocatalysts based on TiO_2–carbon for air pollutant removal: aspects of adsorption. Catal Today. https://doi.org/10.1016/j.cattod.2018.09.027
44. Zhong L, Lee CS, Haghighat F (2012) Adsorption performance of titanium dioxide (TiO_2) coated air filters for volatile organic compounds. J Hazard Mater 243:340–349. https://doi.org/10.1016/j.jhazmat.2012.10.042
45. Matos J, Garcia A, Chovelon J, Ferronato C (2010) Combination of adsorption on activated carbon and oxidative photocatalysis on TiO_2 for gaseous toluene remediation. Open Mater Sci J 4:23–25. https://doi.org/10.2174/1874088X01004020023
46. Šuligoj A, Štangar UL, Tušar NN (2014) Photocatalytic air-cleaning using TiO_2 nanoparticles in porous silica substrate. Chem Pap 68:1265–1272. https://doi.org/10.2478/s11696-014-0553-7
47. Li M, Lu B, Ke QF, Guo YJ, Guo YP (2017) Synergetic effect between adsorption and photodegradation on nanostructured TiO_2/activated carbon fiber felt porous composites for toluene removal. J Hazard Mater 333:88–98. https://doi.org/10.1016/j.jhazmat.2017.03.019
48. Shi JW, Cui HJ, Chen JW, Fu ML, Xu B, Luo HY, Ye ZL (2012) TiO_2/activated carbon fibers photocatalyst: effects of coating procedures on the microstructure, adhesion property, and photocatalytic ability. J Colloid Interface Sci 388:201–208. https://doi.org/10.1016/j.jcis.2012.08.038
49. Ahmadkhaniha R, Izadpanah F, Rastkari N (2017) Reduced graphene oxide-TiO_2 nanocomposite facilitated visible light photodegradation of gaseous toluene. J Environ Prot (Irvine, Calif). 08:591–602. https://doi.org/10.4236/jep.2017.85039
50. Lu Y, Wang D, Ma C, Yang H (2010) The effect of activated carbon adsorption on the photocatalytic removal of formaldehyde. Build Environ 45:615–621. https://doi.org/10.1016/j.buildenv.2009.07.019

51. Kim S, Kim M, Lee HY, Yu JS (2017) Visible light-induced photocatalytic degradation of gas-phase acetaldehyde with platinum/reduced titanium oxide-loaded carbon paper. RSC Adv. 7:50693–50700. https://doi.org/10.1039/c7ra10778a
52. Ouzzine M, Romero-Anaya AJ, Lillo-Ródenas MA, Linares-Solano A (2014) Spherical activated carbon as an enhanced support for TiO_2/AC photocatalysts. Carbon 67:104–118. https://doi.org/10.1016/j.carbon.2013.09.069
53. Mull B, Möhlmann L, Wilke O (2017) Photocatalytic degradation of toluene, butyl acetate and limonene under UV and visible light with titanium dioxide-graphene oxide as photocatalyst. Environments 4:9. https://doi.org/10.3390/environments4010009
54. Aghighi A, Haghighat F (2015) Evaluation of nano-titanium dioxide (TiO_2) catalysts for ultraviolet photocatalytic oxidation air cleaning devices. J Environ Chem Eng 3:1622–1629. https://doi.org/10.1016/j.jece.2015.05.019
55. Han Z, Chang VWC, Zhang L, Tse MS, Tan OK, Hildemann LM (2012) Preparation of TiO_2-coated polyester fiber filter by spray-coating and its photocatalytic degradation of gaseous formaldehyde. Aerosol Air Qual. Res. 12:1327–1335. https://doi.org/10.4209/aaqr.2012.05.0114
56. Boccaccini AR, Karapappas P, Marijuan JM, Kaya C (2004) Electrophoretic deposition : fundamentals and applications TiO_2 coatings on silicon carbide and carbon fiber substrates by electrophoretic deposition. 9:851–859
57. Sánchez B, Sánchez-Muñoz M, Muñoz-Vicente M, Cobas G, Portela R, Suárez S, González AE, Rodríguez N, Amils R (2012) Photocatalytic elimination of indoor air biological and chemical pollution in realistic conditions. Chemosphere 87:625–630. https://doi.org/10.1016/j.chemosphere.2012.01.050
58. Ding Z, Hu X, Lu GQ, Yue P-L, Greenfield PF (2000) Novel silica gel supported TiO_2 photocatalyst synthesized by CVD method. Langmuir 16:6216–6222. https://doi.org/10.1021/la000119l
59. Raza N, Kim KH, Agbe H, Kailasa SK, Szulejko JE, Brown RJC (2017) Recent advances in titania-based composites for photocatalytic degradation of indoor volatile organic compounds. Asian J Atmos Environ 11:217–234. https://doi.org/10.5572/ajae.2017.11.4.217
60. Tian MJ, Liao F, Ke QF, Guo YJ, Guo YP (2017) Synergetic effect of titanium dioxide ultralong nanofibers and activated carbon fibers on adsorption and photodegradation of toluene. Chem Eng J 328:962–976. https://doi.org/10.1016/j.cej.2017.07.109
61. Lekshmi MV, Shiva Nagendra SM, Maiya MP (2019) Photocatalytic degradation of gaseous toluene using self-assembled air filter based on chitosan/activated carbon/TiO_2. Journal of Environmental Chemical Engineering 7(6):103455. https://doi.org/10.1016/j.jece.2019.103455
62. Héquet V, Raillard C, Debono O, Thévenet F, Locoge N, Le Coq L (2018) Photocatalytic oxidation of VOCs at ppb level using a closed-loop reactor: the mixture effect. Appl Catal B Environ 226:473–486. https://doi.org/10.1016/j.apcatb.2017.12.041
63. Kolarik B, Wargocki P, Skorek-Osikowska A, Wisthaler A (2010) The effect of a photocatalytic air purifier on indoor air quality quantified using different measuring methods. Build Environ 45:1434–1440. https://doi.org/10.1016/j.buildenv.2009.12.006
64. Gunschera J, Markewitz D, Bansen B, Salthammer T, Ding H (2016) Portable photocatalytic air cleaners: efficiencies and by-product generation. Environ Sci Pollut Res 23:7482–7493. https://doi.org/10.1007/s11356-015-5992-3
65. Yu KP, Whei-May Lee G, Huang WM, Wu CC, Lou CI, Yang S (2006) Effectiveness of photocatalytic filter for removing volatile organic compounds in the heating, ventilation, and air conditioning system. J Air Waste Manag Assoc 56:666–674. https://doi.org/10.1080/10473289.2006.10464482
66. Ramirez AM, Demeestere K, De Belie N, Mäntylä T, Levänen E (2010) Titanium dioxide coated cementitious materials for air purifying purposes: preparation, characterization and toluene removal potential. Build Environ 45:832–838. https://doi.org/10.1016/j.buildenv.2009.09.003
67. Negishi N, Matsuzawa S, Takeuchi K, Pichat P (2007) Transparent micrometer-thick TiO_2 films on SiO_2-coated glass prepared by repeated dip-coating/calcination: characteristics and photocatalytic activities for removing acetaldehyde or toluene in air. Chem Mater 19:3808–3814. https://doi.org/10.1021/cm070320i

68. Lorencik S, Yu QL, Brouwers HJH (2016) Photocatalytic coating for indoor air purification: synergetic effect of photocatalyst dosage and silica modification. 306:942–952. https://doi.org/10.1016/j.cej.2016.07.093
69. Auvinen J, Wirtanen L (2008) The influence of photocatalytic interior paints on indoor air quality. Atmos Environ 42:4101–4112. https://doi.org/10.1016/j.atmosenv.2008.01.031
70. Maggos T, Bartzis JG, Liakou M, Gobin C (2007) Photocatalytic degradation of NO_x gases using TiO_2-containing paint: a real scale study. J Hazard Mater 146:668–673. https://doi.org/10.1016/j.jhazmat.2007.04.079
71. Kolarik J, Toftum J (2012) The impact of a photocatalytic paint on indoor air pollutants: sensory assessments. Build Environ 57:396–402. https://doi.org/10.1016/j.buildenv.2012.06.010
72. Mamaghani AH, Haghighat F, Lee CS (2017) Photocatalytic oxidation technology for indoor environment air purification: the state-of-the-art. Appl Catal B Environ 203:247–269. https://doi.org/10.1016/j.apcatb.2016.10.037
73. Korologos CA, Philippopoulos CJ, Poulopoulos SG (2011) The effect of water presence on the photocatalytic oxidation of benzene, toluene, ethylbenzene and m-xylene in the gas-phase. Atmos Environ 45:7089–7095. https://doi.org/10.1016/j.atmosenv.2011.09.038
74. Bouazza N, Lillo-Ródenas MA, Linares-Solano A (2008) Photocatalytic activity of TiO_2-based materials for the oxidation of propene and benzene at low concentration in presence of humidity. Appl Catal B Environ 84:691–698. https://doi.org/10.1016/j.apcatb.2008.06.002
75. Mo J, Zhang Y, Xu Q (2013) Effect of water vapor on the by-products and decomposition rate of ppb-level toluene by photocatalytic oxidation. Appl Catal B Environ 132–133:212–218. https://doi.org/10.1016/j.apcatb.2012.12.001
76. Biard PF, Bouzaza A, Wolbert D (2007) Photocatalytic degradation of two volatile fatty acids in an annular plug-flow reactor; kinetic modeling and contribution of mass transfer rate. Environ Sci Technol 41:2908–2914. https://doi.org/10.1021/es062368n
77. Sleiman M, Conchon P, Ferronato C, Chovelon JM (2009) Photocatalytic oxidation of toluene at indoor air levels (ppbv): towards a better assessment of conversion, reaction intermediates and mineralization. Appl Catal B Environ 86:159–165. https://doi.org/10.1016/j.apcatb.2008.08.003
78. Zhu X, Chang D, Li X, Sun Z, Deng X, Zhu A (2015) Inherent rate constants and humidity impact factors of anatase TiO_2 film in photocatalytic removal of formaldehyde from air. Chem Eng J 279:897–903. https://doi.org/10.1016/j.cej.2015.05.095
79. Zhong L, Haghighat F, Lee CS, Lakdawala N (2013) Performance of ultraviolet photocatalytic oxidation for indoor air applications: systematic experimental evaluation. J Hazard Mater 261:130–138. https://doi.org/10.1016/j.jhazmat.2013.07.014
80. Zhong L, Lee CS, Haghighat F, Bahloul A (2016) Deactivation and ultraviolet C-induced regeneration of photocatalytic oxidation air filters. Sci Technol Built Environ 22:576–585. https://doi.org/10.1080/23744731.2016.1171629
81. Huang H, Liu G, Zhan Y, Xu Y, Lu H, Huang H (2017) Photocatalytic oxidation of gaseous benzene under VUV irradiation over TiO_2/zeolites catalysts. Catal Today 281:649–655. https://doi.org/10.1016/j.cattod.2016.07.005
82. Taoda H, Fukaya M, Watanabe E, Tanaka K (2009) VOC decomposition by photocatalytic wall paper. Mater Sci Forum 510–511:22–25. https://doi.org/10.4028/www.scientific.net/msf.510-511.22
83. Chen DH, Ye XJ, Li KY (2005) Oxidation of PCE with a UV LED photocatalytic reactor. Chem Eng Technol 28:95–97. https://doi.org/10.1002/ceat.200407012
84. Lin L, Chai Y, Zhao B, Wei W, He D, He B, Tang Q (2013) Photocatalytic oxidation for degradation of VOCs. Open J Inorg Chem 3:14–25. https://doi.org/10.4236/ojic.2013.31003
85. Ao CH, Lee SC (2004) Combination effect of activated carbon with TiO_2 for the photodegradation of binary pollutants at typical indoor air level. J Photochem Photobiol A Chem 161:131–140. https://doi.org/10.1016/S1010-6030(03)00276-4
86. Luo Y, Ollis DF (1996) Heterogeneous photocatalytic oxidation of trichloroethylene and toluene mixtures in air: Kinetic promotion and inhibition, time-dependent catalyst activity. J Catal 163:1–11. https://doi.org/10.1006/jcat.1996.0299

87. Zhong L, Haghighat F (2015) Photocatalytic air cleaners and materials technologies—abilities and limitations. Build Environ 91:191–203. https://doi.org/10.1016/j.buildenv.2015.01.033
88. Dumont É, Héquet V (2017) Determination of the Clean Air Delivery Rate (CADR) of photocatalytic oxidation (PCO) purifiers for indoor air pollutants using a closed-loop reactor. Part I: Theoretical considerations, Molecules. 22. https://doi.org/10.3390/molecules22030407
89. Zhong L, Haghighat F, Lee CS (2013) Ultraviolet photocatalytic oxidation for indoor environment applications: experimental validation of the model. Build Environ 62:155–166. https://doi.org/10.1016/j.buildenv.2013.01.009
90. Hossain M, Raupp GB, Hay SO, Obee TN (1999) Three-dimensional developing flow model for photocatalytic monolith reactors, 45
91. Zhong L, Haghighat F (2011) Modeling and validation of a photocatalytic oxidation reactor for indoor environment applications. Chem Eng Sci 66:5945–5954. https://doi.org/10.1016/j.ces.2011.08.017
92. Imoberdorf GE, Taghipour F, Keshmiri M, Mohseni M (2008) Predictive radiation field modeling for fluidized bed photocatalytic reactors. 63:4228–4238. https://doi.org/10.1016/j.ces.2008.05.022
93. Puma GL (2007) Radiation field optimization in photocatalytic monolith reactors for air treatment, 53. https://doi.org/10.1002/aic
94. Verbruggen SW, Keulemans M, van Walsem J, Tytgat T, Lenaerts S, Denys S (2016) CFD modeling of transient adsorption/desorption behavior in a gas phase photocatalytic fiber reactor. Chem Eng J 292:42–50. https://doi.org/10.1016/j.cej.2016.02.014
95. Roegiers J, van Walsem J, Denys S (2018) CFD- and radiation field modeling of a gas phase photocatalytic multi-tube reactor. Chem Eng J 338:287–299. https://doi.org/10.1016/j.cej.2018.01.047
96. Raupp GB, Alexiadis A, Hossain M, Changrani R (2001) First-principles modeling, scaling laws and design of structured photocatalytic oxidation reactors for air purification. 69:41–49
97. Den W, Wang CC (2012) Enhancement of adsorptive chemical filters via titania photocatalysts to remove vapor-phase toluene and isopropanol. Sep Purif Technol 85:101–111. https://doi.org/10.1016/j.seppur.2011.09.054

Experimental Investigation of ISHRAE IEQ Standard Focusing on Implementation Aspects Through Pilot Study

Shailendra Kumar, Nikhil Jain and Jyotirmay Mathur

Abstract In the year 2016, Indian Society for Heating, Refrigerating and Air-Conditioning Engineers (ISHRAE) released India's first indoor environmental quality (IEQ) standard. The present study demonstrates review of this IEQ standard and shows the effect of seasonal variation on IEQ parameters. Measurements are taken in two buildings having zones such as individual offices, open-plan offices, classrooms, etc. Observations spanning over 12 months revealed that the arithmetic mean value of thermal comfort parameters, indoor operative temperature, relative humidity (RH) and floor surface temperature, are 26.6 ± 3.9 °C, 46 ± 17% and 26.7 ± 4.5 °C, respectively. Indoor air quality (IAQ) parameters, CO_2, $PM_{2.5}$ and PM_{10} concentrations, are 570 ± 176 ppm, 46 ± 22 $\mu g/m^3$ and 116 ± 37 $\mu g/m^3$, respectively. Lighting comfort parameters, illuminance and circadian lighting design, are 326 ± 140 lx and 341 ± 146 EML, respectively. Occupant satisfaction survey is also conducted, and it is observed that occupant satisfaction for indoor air quality, thermal and lighting comfort is 80%, 81% and 89%, respectively. Majority of the IEQ parameters in both the buildings are found to be meeting the minimum threshold.

Keywords Indoor environmental quality · Field study · Standard

1 Introduction

Healthy and comfortable indoor environment is primary requirement of the building occupants as in urban environment residents spent most of their time indoors [8]. IEQ is affected by building location, orientation, climatic conditions, occupant behaviour and building systems and typology [15]. Indoor air pollutants have clear association with the sense of well-being, occupant health and productivity. In addition to long-term, they can worsen conditions such as asthma, headache, respiratory diseases, allergies, etc. Furthermore, both too high or too low temperature and humidity, excessively bright or deficient lighting system, high noise level leads to increase

S. Kumar · N. Jain · J. Mathur (✉)
Centre for Energy and Environment, Malaviya National Institute of Technology, Jaipur, India
e-mail: jmathur.mech@mnit.ac.in

A. Sharma et al. (eds.), *Indoor Environmental Quality*,
Lecture Notes in Civil Engineering 60,
https://doi.org/10.1007/978-981-15-1334-3_17

in mental and physical stress levels in human body leading to health problems such as concentration impairment, memory loss, digestive problem, sleep disorder, etc. [9, 10]. Thermal comfort can be defined as "the state of mind which expresses the satisfaction with thermal environment" [1]. It relates to the physical factors in air-conditioned as well as naturally ventilated environment and therefore strongly correlated with building energy consumption [4].

Improper lighting conditions such as too bright or too dark can create visual discomfort. Lighting comfort directly affects the occupant's work productivity and efficiency. It is a subjective measure and directly dependent on factors like illumination, the risk of glare, luminous spectrum, brightness and luminance [6]. A proper visual environment condition causes an increment in the productivity and well-being of the building occupants [16].

IAQ is recognised to cause chronic and acute effects on the building occupant's health. It is directly connected to the concentration of pollutants and ventilation rates, which is directly linked to Sick Building Syndrome [18]. For indoor environment, indoor air quality is associated to chemical as well as physical causes which are concentrations of CO_2, CO, respirable suspended particulate matter ($PM_{2.5}$, PM_{10}), total volatile organic compounds (TVOC), formaldehyde (CH_2O), SO_2, NO_2, etc.

In the year 2016, ISHRAE released India's first IEQ standard [11]. In this standard, thermal comfort, indoor air quality, acoustic comfort and lighting comfort are defined as elements of IEQ, and they have been further divided into parameters as shown in Fig. 1.

Standard includes definitions for the elements affecting health and comfort, threshold values of parameters contributing to these elements, measurement methodology,

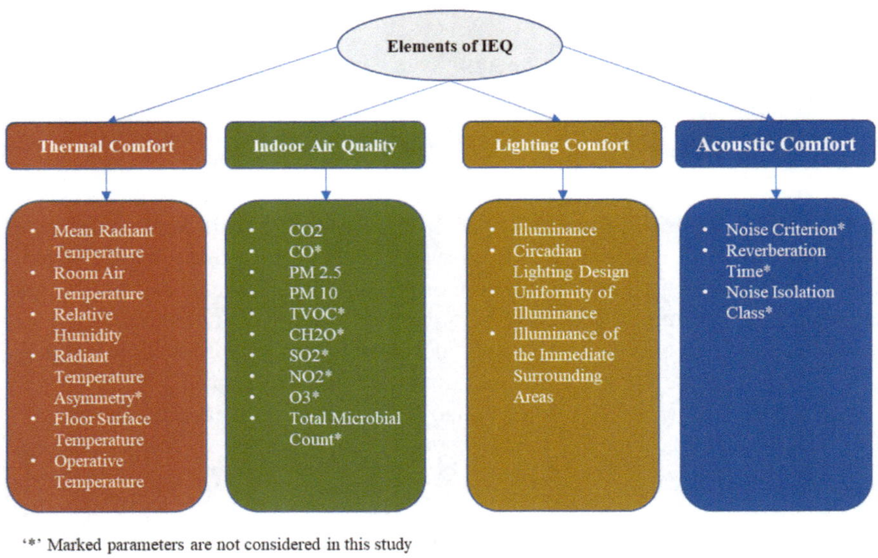

Fig. 1 IEQ elements and parameter affecting these elements

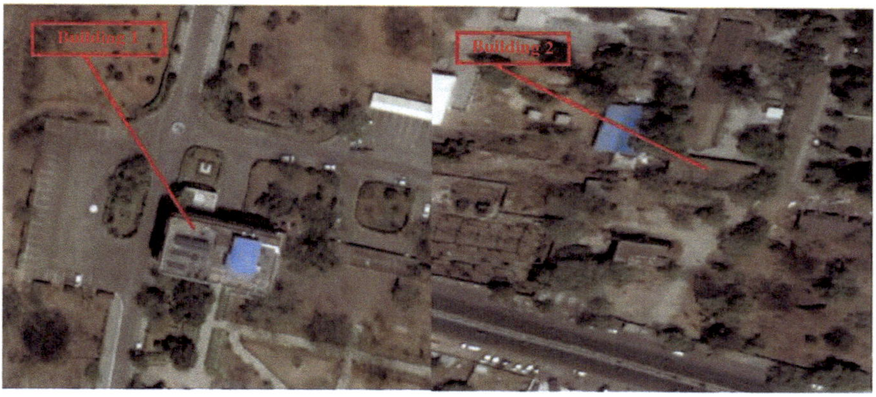

Fig. 2 Geographical location of Building 1 and Building 2. *Source* Google Maps: https://www.google.com/maps/place/Malaviya+National+Institute+of+Technology+Jaipur/@26.8617377,75.8091142,824m/data=!3m1!1e3!4m5!3m4!1s0x396db66fe2879c7f:0xdfc843bf9b6f869a!8m2!3d26.8630144!4d75.810592

specifications of measurement instruments and satisfaction survey. The threshold values of parameters have been defined as three levels such as Class A (aspirational), Class B (acceptable) and Class C (marginally acceptable). Class A level is comparable to international standards. As this standard is applicable to residential as well as commercial buildings which can have conditioned as well as unconditioned spaces, some of the parameters are excluded in Class B and Class C.

In this study, measurement methodology proposed in ISHRAE's IEQ standard is reviewed and analysed through a pilot testing on two buildings in terms of where, when and how to take measurements. Also, effect of seasonal variations on IEQ parameter are analysed. Furthermore, along with physical parameter measurements, occupant satisfaction surveys are also conducted.

2 Methodology

2.1 *Overall Approach*

In this study, IEQ measurements and survey are done three times a day and twice a season at every potential measurement location. Time slot for measurement and survey during a day is morning session (10.00 a.m. to 11.00 a.m.), afternoon session (2.00 p.m. to 3.00 p.m.) and evening session (4.00 p.m. to 5.00 p.m.).

2.2 Site Description

The study is carried out at Malaviya National Institute of Technology, Jaipur (26.8630° N, 75.8106° E). Jaipur is categorised in composite climate zone. Four seasons are considered in composite climate zone. Summer seasons start from April and last till June, July and August are considered as monsoon season, September as post-monsoon season (considered as summer for present study), November to February as winter season, whereas, October and March are moderate seasons [13]. Two buildings which are part of educational institutional have been selected for IEQ investigations. Buildings are occupied between 8 a.m. and 5 p.m. during weekdays. The detailed descriptions of both buildings are shown in Table 1.

The first building named hereafter as 'Building 1' is located away from heavy traffic and is situated in an area with rich natural vegetation, while second building named hereafter as 'Building 2' is only 65 m away from the roadway. Both buildings have shared as well as individual offices, classrooms, conference halls and waiting rooms in which some are air-conditioned, and others are unconditioned. Further, pictorial views and geographical locations of Building 1 and Building 2 are shown in Figs. 3 and 2, respectively.

Table 1 Building description

Parameters	Building 1	Building 2
Area (Sq. Metre)	1400 (distributed over three floors having 25 rooms)	165 (single floor and three rooms)
Construction materials	Stone walls (Ground floor), Brick walls (First and Second floor), RCC roofing	Stone walls, asbestos roofing, acoustic ceiling tiles
Orientation	South-East, North-West	South-East, North-West
Ventilation type	Mixed mode	Mixed mode
Occupancy	75 and variable	13 and variable
Measured occupant locations	66	10

Fig. 3 Pictorial view of Building 1 and Building 2

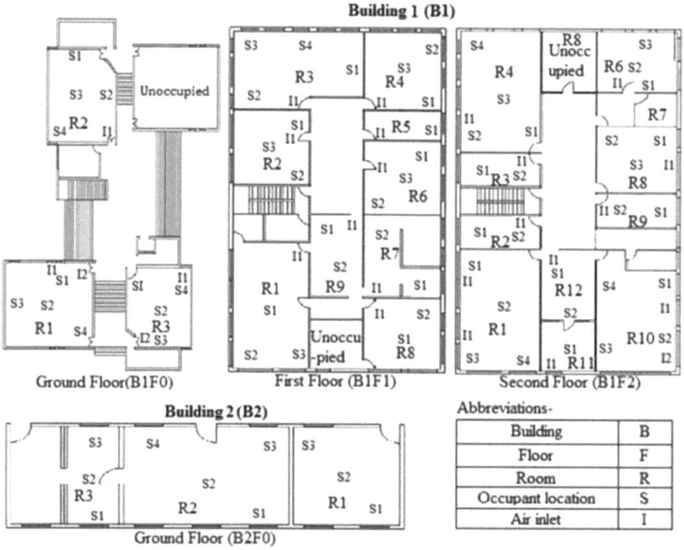

Fig. 4 Floor plan and potential occupant location at old administration and CEE building

3 Potential Measurement Locations

All parameters are measured at different potential measurement locations in the identified buildings. Potential measurement locations are those where extreme value of IEQ parameters can be observed such as near and away to the closed and open window, door, ventilation system, artificial light source, under ceiling fan and air-conditioning unit. Also, centre of the room is also considered potential measurement location to cover overall effect within specified space [7, 11]. All locations are marked in Fig. 4. Both occupant satisfaction survey and IEQ parameter measurements are recorded at every potential measurement location. In this study, total 1824 occupant responses are collected which include responses of 1520 males and 204 females. This has done for four seasons spanning June 2017 and May 2018.

4 Instruments

Measurements are done by the instruments meeting the accuracy and resolution criteria specified in IEQ standard [11]. The instruments are procured from the standard manufacturers, and the details of the instruments used are given in Table 2.

Table 2 Specifications of instrument used

Instrument name (manufacturer name-model)	Parameter	Resolution	Accuracy	Measurement methodology/sensor type
Thermal comfort analyser (Testo-480)	Air temperature	0.1 °C	± 0.5 °C	Thermistor effect of metals
	Mean radiant temperature	0.1 °C	±0.5 °C	Thermocouple type K
	Relative humidity	1%	±3%	
	Air velocity	0.01 m/s	±0.05 m/s	Hot wire anemometer
	CO_2	1 ppm	±5%	NDIR
	Illuminance	1 lx	±5%	V-lambda curve
Surface thermometer (Testo 905-T2)	Floor surface temperature	0.5 °C	±1 °C	Contact thermometer
Air quality monitor	$PM_{2.5}$	1 $\mu g/m^3$	±5%	Light scattering
(ATS-Y09-PM)	PM_{10}	1 $\mu g/m^3$	±5%	Light scattering

5 Measurement Methodology

All IEQ parameters are measured and analysed as specified in ISHRAE IEQ standard [11]. Parameters such as mean radiant temperature, room air temperature, air velocity and relative humidity representing thermal comfort are recorded at 0.6 m and 1.1 m above floor for seating and standing positions, respectively [11, 14]. Operative temperature can be obtained as the average value of room air temperature and mean radiant temperature for occupants involved in sedentary physical activity (1.0 < met < 1.3) provided the occupant is not exposed to direct sunlight and for air velocities lower than 0.2 m/s. For air velocity up to 0.2 m/s, the acceptable range of operative temperature is given in ISHRAE standard [11]. Higher air velocity attributed towards an increase in comfortable operative temperature, the required acceptable range can be obtained by using approach given in ISO 7730 [12], and the same is suggested in ISHRAE IEQ standard [11]. IAQ parameters, CO_2, $PM_{2.5}$ and PM_{10}, are measured at occupant locations and at possible fresh air intake locations in an occupant zone. CO_2 concentration is measured when there is at least 90% occupancy. As defined in IEQ standard, illuminance of task area is calculated by taking arithmetic mean values of illuminance at different points within the task area. Equivalent melanopic lux (EML) is calculated as combined effect of the illuminance and melanopic ratio. Melanopic ratio is dependent on specifications of light source such as correlated colour temperature and light source type of light source. Uniformity of illuminance is calculated as the ratio of minimum illuminance to the average illuminance of different points within considered the task area [5, 11].

6 Results and Discussions

6.1 Thermal Comfort

In this section, seasonal influence on parameters of thermal comfort such as operative temperature, air velocity, relative humidity and floor surface temperature is analysed.

Operative temperature measures the effects of radiation and convection heat transfers in actual nonuniform indoor climate [1]. Seasonal variations in operative temperature and air velocity are presented in Figs. 5 and 6, respectively. During pre-monsoon, ambient temperature and indoor operative temperature for the month of June are observed quite low as compared to other months of summer season. For the winter season, operative temperature is found within comfortable range without using artificial heating source. Further, in summer and monsoon season, ceiling fans are used for cooling purpose in naturally ventilated as well as at times in air-conditioned rooms. However, due to an increase in air velocity, some of the potential occupant locations are found in the acceptable range even at a higher operative temperature up to 30.5 °C.

Since Jaipur lies in composite climate where RH varies substantially across the year and observed to be 32%, 59%, 34% and 28%, in summer, monsoon, winter and moderate season, respectively. Most data obtained for RH falls within as ISHRAE IEQ standard acceptable range. However, discomfort due to low RH arises because of high-temperature conditions around the year. Moreover, the occupant satisfaction level achieved for threshold range of RH is found to be 80%. Seasonal variations

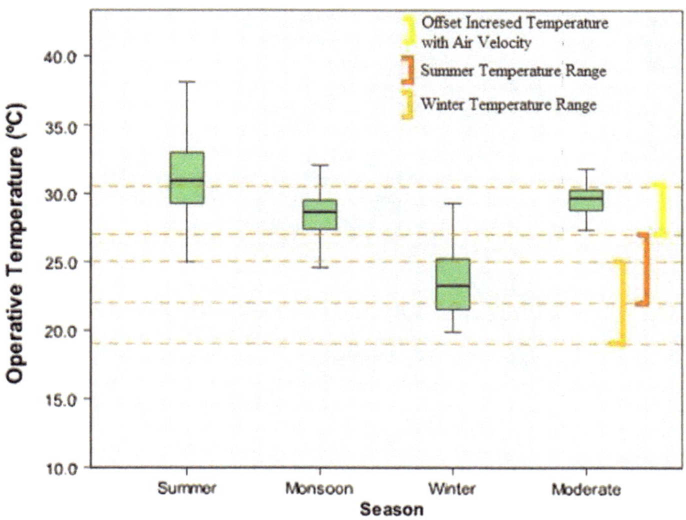

Fig. 5 Seasonal variations in operative temperature

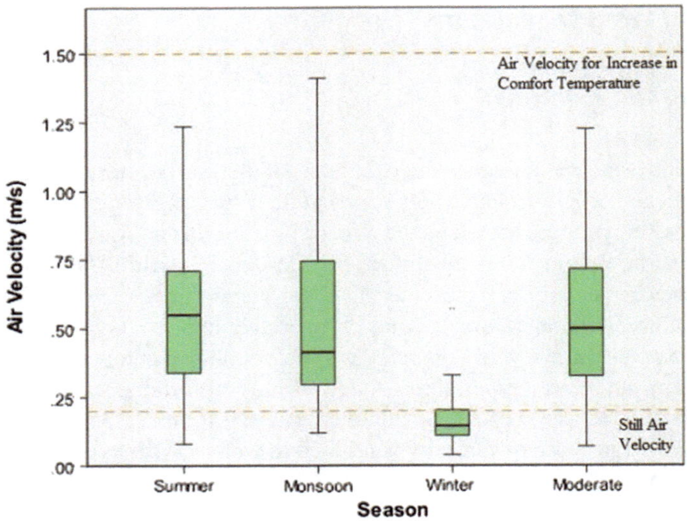

Fig. 6 Seasonal variations in air velocity

in relative humidity and floor surface temperature are presented in Figs. 7 and 8, respectively.

Floor surface temperature is measured as occupant may feel discomfort because of too warm or too cool temperature of floor. As per ISHRAE IEQ standard, floor surface

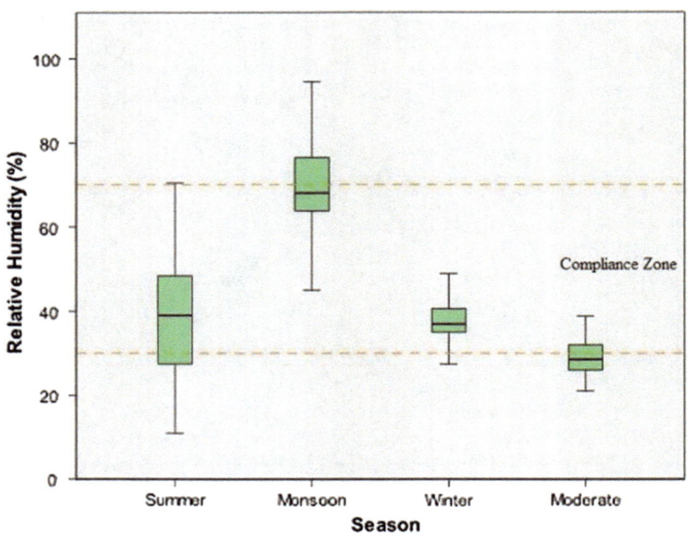

Fig. 7 Seasonal variations in relative humidity

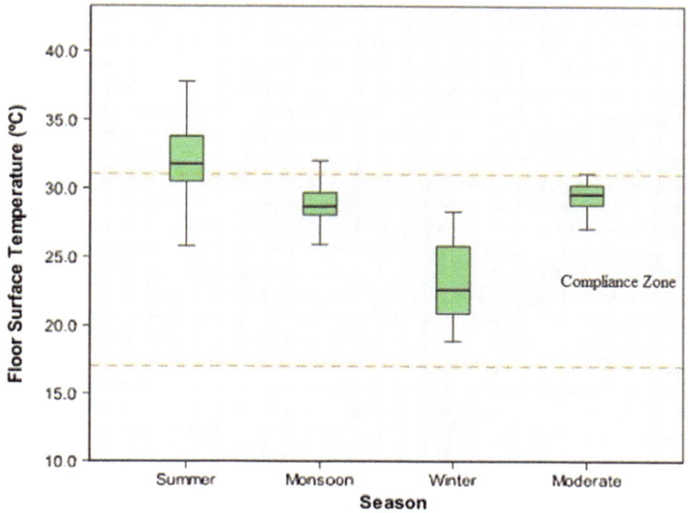

Fig. 8 Seasonal variations in floor surface temperature

temperature for the occupants heaving feet contacts with floor shall be 17–31 °C within the occupant zone. In this study, the mean value of floor surface temperature for summer, monsoon, winter and moderate seasons is found to be 33.4 °C, 29.7 °C, 21.3 °C and 28.2 °C, respectively. During summer, the floor surface temperature exceeds the comfort limit since the ambient temperature in the aforesaid location hits more than 40 °C.

6.2 *Indoor Air Quality*

Another important element of IEQ is indoor air quality which specifies the indoor air characteristics in terms of concentrations of CO_2, $PM_{2.5}$ and PM_{10}. The details of each have been discussed in this section.

The concentration of CO_2 is measured at each potential occupant locations in the room with minimum 90% occupancy. Due to high natural vegetation in MNIT campus, the ambient CO_2 concentration is found to be 400 ± 25 ppm. The seasonal variations in CO_2 concentration are illustrated in Fig. 9. In the present study, the mean value of CO_2 concentrations for summer, monsoon, winter and moderate seasons are found to be 534 ppm, 443 ppm, 512 ppm and 479 ppm, respectively. The CO_2 concentration range for all seasons is falling under Class A, as specified by ISHRAE IEQ standard, and insignificant seasonal influence is observed. Higher CO_2 concentration is recorded in the classrooms and split air-conditioned rooms which are due to high occupancy [17] and less fresh air intake, respectively.

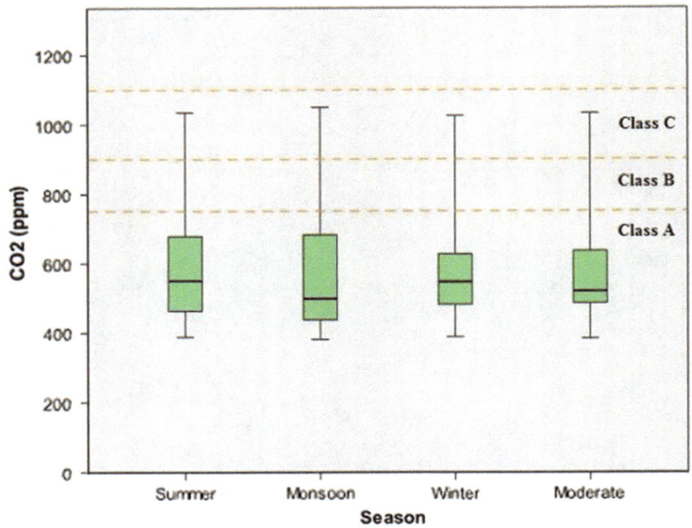

Fig. 9 Seasonal variations in CO_2 concentration

The presence of particulate matter (PM) in environment affects the indoor air quality to large extent. The seasonal variations in $PM_{2.5}$ and PM_{10} concentration are illustrated in Figs. 10 and 11, respectively. The mean value of $PM_{2.5}$ concentrations for summer, monsoon, winter and moderate seasons is found to be 27 $\mu g/m^3$,

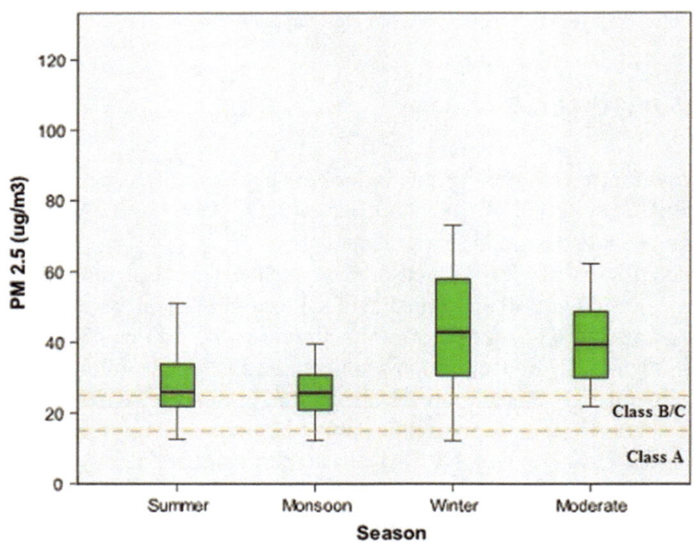

Fig. 10 Seasonal variations in $PM_{2.5}$ concentration

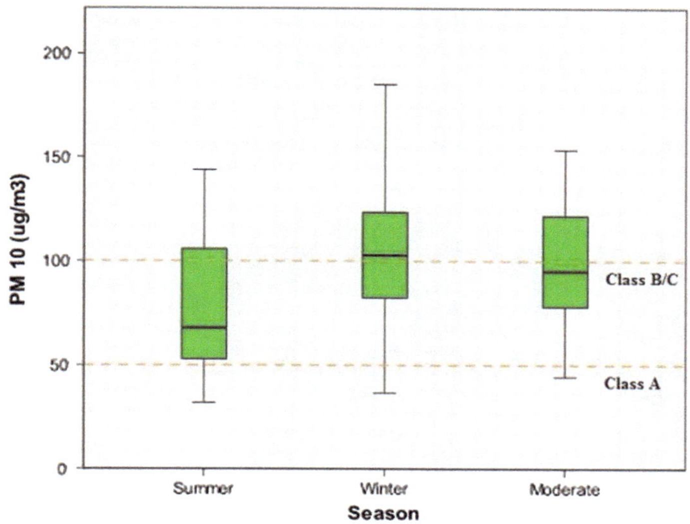

Fig. 11 Seasonal variations in PM_{10} concentration

25 μg/m³, 61 μg/m³ and 44 μg/m³, respectively. Further, the mean value of PM_{10} concentration for summer, winter and moderate seasons is found to be 68 μg/m³, 107 μg/m³ and 89 μg/m³, respectively. In monsoon season, the lower concentrations of particulate matters are observed due to washout of the particles from the local environment. In the summer season, lower particulate matter concentrations are observed because of better dispersion of particulates due to high temperature, hot air to move upwards. During winter, the particulate matter concentrations are recorded high due to less dispersion caused by lower temperature [2].

6.3 Lighting Comfort

Lighting is another IEQ element which influences many bodily functions like the nervous system, circadian rhythms, pituitary gland, endocrine system, pineal gland and alertness due to its different wavelength [11]. Illuminance and circadian lighting design are interrelated with each other. The equivalent melanopic lux (EML) is the measurement of circadian lighting design and can be calculated by multiplying the illuminance (Lux) and melanopic ratio. The EML expresses the biological effects of light on the human body [11]. In the present study, the mean value of illuminance for summer, monsoon, winter and moderate seasons is found to be 286 lx, 310 lx, 346 lx and 316 lx, respectively. Further, the mean value of circadian lighting design for summer, monsoon, winter and moderate seasons is found to be 304 EML, 328 EML, 365 EML and 334 EML, respectively.

Seasonal variations of illuminance and circadian lighting design are depicted in Figs. 12 and 13, respectively. During monsoon season, relatively low mean values of illuminance and corresponding EML are observed because of cloudy sky. In winter, relatively high values of illuminance and EML are observed because of occupants open blinds and curtains to gain more solar heat leading to more daylight in occupant

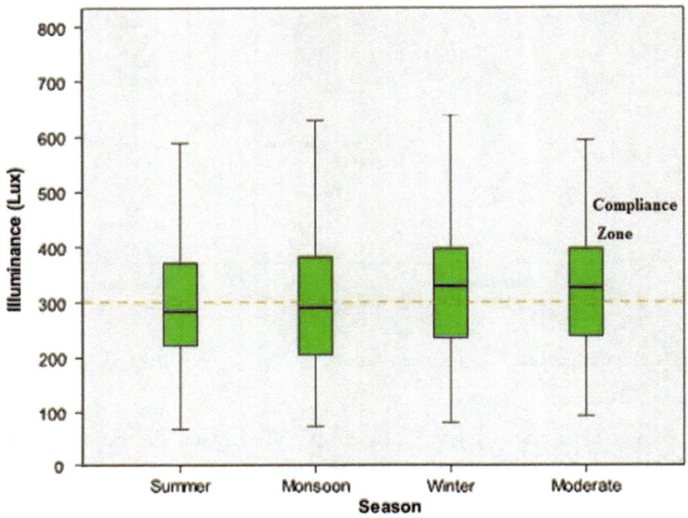

Fig. 12 Seasonal variations of illuminance

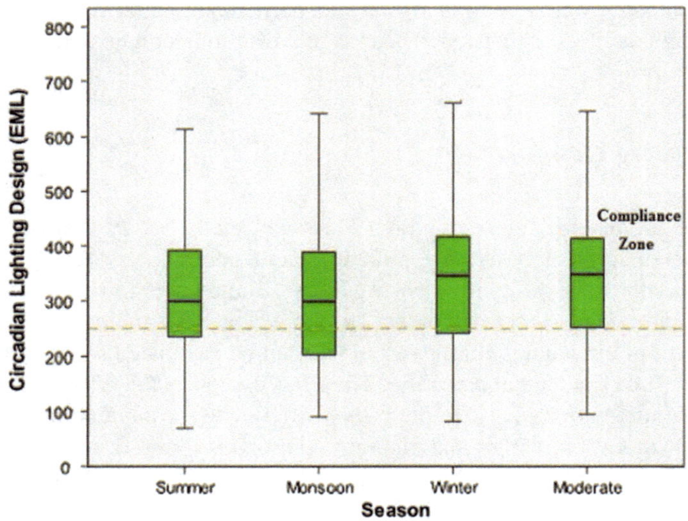

Fig. 13 Seasonal variations of circadian lighting design

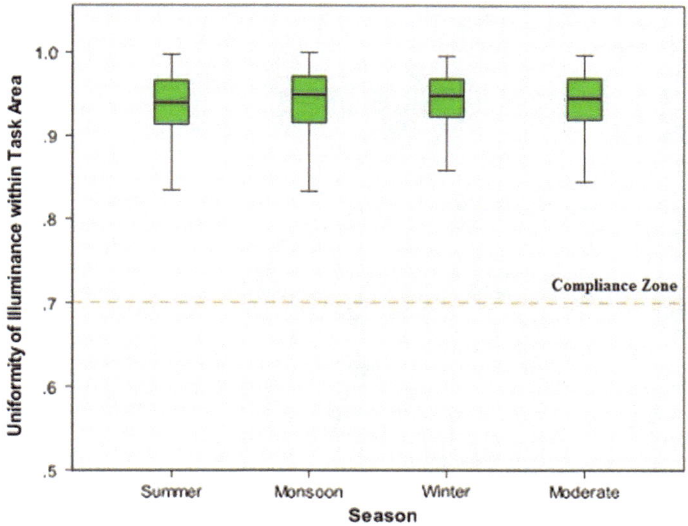

Fig. 14 Seasonal variations of uniformity of illuminance within task area

zone. In summer season, lowest values of illuminance and EML are observed because of occupant tendency to close blinds and curtains to avoid solar heat transmission into the space leading to less daylight in occupant zone.

The distribution of light across the task area should be uniform for better visual comfort [5]. The ratio of illuminance available in task area to the illuminance available in immediate adjacent surrounding should be within the threshold limits as per standard. Seasonal variations of same are shown in graphs as shown in Figs. 14 and 15. Both uniformity of illuminance and ratio of illuminance of task area to immediately adjacent surroundings are in acceptable range. However, occupant location near to East and West facing windows get direct sunlight leads both parameters (uniformity of illuminance and ratio of illuminance in task area to immediate adjacent surrounding) in the unacceptable range.

6.4 Occupant Satisfaction Surveys

In present study, subjective assessment of IEQ elements (thermal comfort, lighting comfort and indoor air quality) is conducted using anonymous occupant satisfaction survey, and responses are recorded on a 7-point scale [11]. Occupant satisfaction surveys are conducted at every potential occupant location simultaneously with IEQ parameter measurements. On 7-point scale, 1 and 7 represent the most unsatisfactory and satisfactory conditions, respectively. Each parameter is represented in terms of percentage satisfaction, where responses 4, 5, 6 and 7 are taken as satisfactory and 1, 2

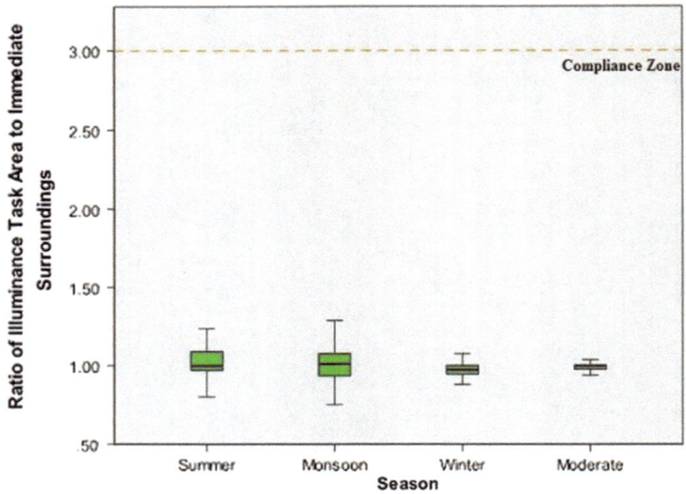

Fig. 15 Seasonal variations of ratio of illuminance task area to immediate surroundings

and 3 as unsatisfactory responses. Gender, weight, age, height and native place of the respondents are also recorded in the present investigation as suggested in study given in [3]. Qualitative aspects of parameters such as room temperature, RH, airflow, stale air, overall lighting, an external view and daylight availability are recorded in the questionnaire. The threshold values for occupant responses are 90 and 80 percentage for Class A and Class B, respectively. The occupant satisfaction for indoor air quality, thermal and lighting comfort is found to be 80, 81 and 89 percentage, respectively.

7 Conclusions

The present study shows importance of deploying efficient methodology to capture spatial-temporal variation in IEQ elements and parameters for any given potential measurement zone. Present study reflects the results for the given location and buildings. It is suggested that built environment used for other purposes should also evaluate implementation of this standard as per their requirement. Measurement locations, frequency and methodology presented in this study allow to gauge IEQ in studied buildings satisfactorily. For two buildings selected for this study, most of the thermal comfort parameters are found to be meeting the minimum threshold. The acceptable range of comfort operative temperature can be increased at elevated air velocity using fan, which is relatively lower energy-consuming method to achieve thermal comfort. CO_2 concentration is found to be within acceptable range because of natural vegetation in vicinity. The concentration of respirable suspended particulate matter is found to be high from October to January because of surface temperature inversion in winters. Further, because of adequate natural and artificial light, majority of lighting comfort parameters are found to be within acceptable range. The collection of occupant satisfaction survey along with physical parameter measurement is the most effective way for IEQ assessment. During occupant satisfaction survey, majority of occupant responses are found to be satisfying more than 80% for all IEQ elements.

Acknowledgements We wish to thank ISHRAE for providing financial support and members of Centre for Energy and Environment, Malaviya National Institute of Technology, Jaipur, for their support during this study.

References

1. ASHRAE Standard 55:2013: Thermal environment conditions for human occupancy. American Society of Heating, Refrigerating and Air-Conditioning Engineers (2013)
2. Chithra VS, Shiva Nagendra SM (2012) Indoor air quality investigations in a naturally ventilated school building located close to an urban roadway in Chennai, India. Build Environ 54:159–167
3. Dhaka S, Mathur J, Wagner A, Agarwal GD, Garg V (2013) Evaluation of thermal environmental conditions and thermal perception at naturally ventilated hostels of undergraduate students in composite climate. Build Environ 66:42–53
4. Djongyang N, Tchinda R, Njomo D (2010) Thermal comfort: a review paper. Renew Sustain Energy Rev 14:2626–2640
5. EN 12464-1: 2011: Light and lighting-Lighting of workplaces-Part 1: Indoor workplaces. European Standard (2011)
6. EN 12665:2011: Light and lighting- Basic terms and criteria for specifying lighting requirements, Brussels. European Committee for Standardization (2011)
7. Fabbri K, Tronchin L (2015) Indoor environmental quality in low energy buildings. Energy Procedia 78:2778–2783
8. Frontczak M, Wargocki P (2011) Literature survey on how different factors influence human comfort in indoor environments. Build Environ 49:922–937
9. Huang L, Zhu Y, Ouyang Q, Cao B (2012) A study on the effects of thermal, luminous, and acoustic environments on indoor environmental comfort in offices. Build Environ 49:304–309
10. ISHRAE: ISHRAE position paper on indoor environmental quality. Indian Society of Heating, Refrigerating and Air-Conditioning Engineers (2015)
11. ISHRAE Standard 100001:2016: Indoor environmental quality standard. Indian Society of Heating, Refrigerating and Air-Conditioning Engineers (2016)
12. ISO7730:2006: Ergonomics of the thermal environment-analytical determination and interpretation of thermal comfort using calculation of the PMV indices and local thermal comfort criteria. 3rd ed. International Standardization Organization (2006)
13. Kumar S, Singh MK, Loftness V, Mathur J, Mathur S (2016) Thermal comfort assessment and characteristics of occupant's behavior in naturally ventilated buildings in composite climate of India. Energy Sustain Dev 33:108–121
14. Raimondo D, Corgnati SP, Olesen BW (2012) Evaluation methods for indoor environmental quality assessment according to EN15251. REHVA European HVAC 49(4):14–19
15. Sakhare VV, Ralegaonkar RV (2014) Indoor environmental quality: review of parameters and assessment models. Architect Sci Rev 57(2):147–154
16. Serghides DK, Chatzinikola CK, Katafygiotou MC (2015) Comparative studies of the occupants' behaviour in a university building during winter and summer time. Int J Sustain Energ 34:528–551

17. Vilčeková S, Kapalo P, Mečiarová L, Burdová EK, Imreczeová V (2017) Investigation of indoor environment quality in classroom—case study. Energy Procedia 190:496–503
18. Zuhaib S, Manton R, Griffin C, Hajdukiewicz M, Keane MM, Goggins J (2018) An Indoor Environmental Quality (IEQ) assessment of a partially-retrofitted university building. Build Environ 139:69–85